An Introduction to Abstract Algebra
(A One Semester Course)

G. Viglino
Ramapo College of New Jersey
April 2018

CONTENTS

Part 1 PRELIMINARIES

1.1	Functions	1
1.2	Principle of Mathematical Induction	13
1.3	The Division Algorithm and Beyond	21
1.4	Equivalence Relations	29

Part 2 GROUPS

2.1	Definitions and Examples	41
2.2	Elementary Properties of Groups	54
2.3	Subgroups	62
2.4	Homomorphisms and Isomorphisms	72
2.5	Symmetric Groups	84
2.6	Normal Subgroups and Factor Groups	92
2.7	Direct Products	103

Part 3 FROM RINGS TO FIELDS

3.1	Definitions and Examples	111
3.2	Homomorphisms, and Quotient Rings	121
3.3	Integral Domains and Fields	128

Appendix A: Check Your Understanding Solutions

Appendix B: Determinants

Appendix C: Answers to Selected Exercises

PREFACE

This text is specifically designed to be used in a one-semester undergraduate abstract algebra course. It consists of three parts:

PART 1 (25% of text).   Lays a foundation for the algebraic construction that follows.

PART 2 (50% of text).   Focuses exclusively on the primary abstract algebra object: the **GROUP**.

PART 3 (25% of text):   Introduces additional algebra objects, including **Rings** and **Fields**.

For our part, we have made every effort to assist you in the journey you are about to take. We did our very best to write a readable book, without compromising mathematical integrity. Along the way, you will encounter numerous *Check Your Understanding* boxes designed to challenge your understanding of each newly-introduced concept. Detailed solutions to each of the Check Your Understanding problems appear in Appendix A, but you should only turn to that appendix after making a valiant effort to solve the given problem on your own, or with others. In the words of Desecrates:

We never understand a thing so well, and make it our own, when we learn it from another, as when we have discovered it for ourselves.

I wish to thank my colleague, Professor Maxim Goldberg-Rugalev, for his invaluable input throughout the development of this text.

Part 1
Preliminaries

§1. FUNCTIONS

We begin by recalling a bit of set notation and some definitions involving sets:

The symbol "$\in$" is read "is contained in or is an element of." In particular;
$$x \in A \Rightarrow x \in B$$
translates to:
If x is in A, then x is in B

DEFINITION 1.1

SET EQUALITY — Two sets A and B are **equal,** written $A = B$ if:
$$x \in A \Rightarrow x \in B \quad \text{and} \quad x \in B \Rightarrow x \in A$$
(or: $x \in A \Leftrightarrow x \in B$)

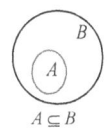
$A \subseteq B$

SUBSET — The set A is said to be a **subset** of the set B, written $A \subseteq B$, if every element in A is also an element in B, i.e: $x \in A \Rightarrow x \in B$.

PROPER SUBSET — A is said to be a **proper subset** of B, written $A \subset B$, if A is a subset of B and $A \neq B$.

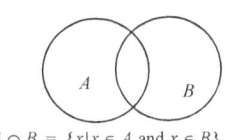
$A \cap B = \{x | x \in A \text{ and } x \in B\}$

INTERSECTION — The **intersection** of A and B, written $A \cap B$, is the set consisting of the elements common to both A and B. That is:
$$A \cap B = \{x | x \in A \text{ and } x \in B\}$$
↖ read *such that*

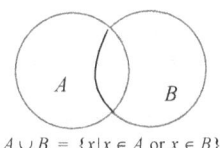
$A \cup B = \{x | x \in A \text{ or } x \in B\}$

UNION — The **union** of A and B, written $A \cup B$, is the set consisting of the elements that are in A or in B (see margin). That is:
$$A \cup B = \{x | x \in A \text{ or } x \in B\}$$

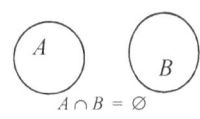
$A \cap B = \emptyset$

DISJOINT SETS — Two sets A and B are **disjoint** if $A \cap B = \emptyset$
the empty set ↗

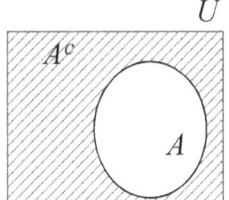

COMPLEMENT — Let A be a subset of the universal set U. The **complement** of A in U, written A^c, is the set of elements in U that are not contained in A:
$$A^c = \{x | x \in U \text{ and } x \notin A\}$$
(More simply: $\{x | x \notin A\}$, if U is understood)

While on the topic of notation we call your attention to the following globally understood mathematical symbols:

$\forall$: read "***for every***" $\exists$: read "***there exists***" $\ni$ read "***such that***"

For example:
$$\forall x \; \exists y \ni x + y > 0$$
is read: for every x there exists y such that $x+y$ is greater than 0

2 Part 1 Preliminaries

All "objects" in mathematics are sets, and functions are no exceptions. The function f given by $f(x) = x^2$, is the subset $f = \{(x, x^2) | x \in \Re\}$ of the plane. Pictorially:

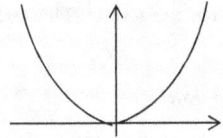

A function such as
$$f = \{(x, x^2) | x \in \Re\}$$
is often simply denoted by $f(x) = x^2$. Still, in spite of their dominance throughout mathematics and the sciences, functions that can be described in terms of algebraic expressions are truly exceptional. Scribble a curve in the plane for which no vertical line cuts the curve in more than one point and you have yourself a function. But what is the "rule" for the set g below?

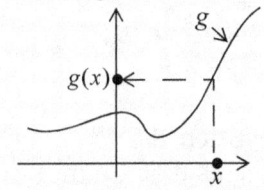

Note that the set S below, is not a function:

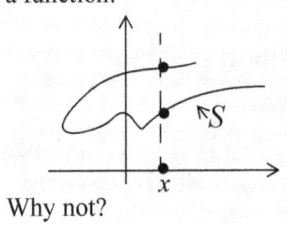

Why not?

You've dealt with functions in one form or another before, but have you ever been exposed to a definition? If so, it probably started off with something like:

A **function** is a rule...........

or, if you prefer, a **rule** is a function.......

You are now too sophisticated to accept this sort of "circular definition." Alright then, have it your way:

DEFINITION 1.2
CARTESIAN PRODUCt

For given sets X and Y, we define the **Cartesian Product** of X with Y, denoted by $X \times Y$, to be the set of **ordered pairs**:
$$X \times Y = \{(x, y) | x \in X \text{ and } y \in Y\}$$

DEFINITION 1.3
FUNCTION

A **function** f from a set X to a set Y is a subset $f \subseteq X \times Y$ such that for every $x \in X$ there exists a <u>unique</u> $y \in Y$.

OPERATOR

A function f from a set X to itself is said to be an **operator** on X.

The symbol $f: X \to Y$ is used to indicate that f is a function from the set X to the set Y, and $y = f(x)$ denotes that $(x, y) \in f$.

DOMAIN

The set X is said to be the **domain** of f, and
$$\{y \in Y | (x, y) \in f \text{ for some } x \in X\}$$

RANGE

is said to be the **range** of f.

While the domain of f is all of X, the range of f need not be all of Y.

Moreover, for $A \subseteq X$ and $B \subseteq Y$:

IMAGE OF $A \subseteq X$

$f[A] = \{f(a) | a \in A\}$ is called the **image of A under f**, and $f^{-1}[B] = \{x \in X | f(x) \in B\}$ is called the **inverse image of B**.

INVERSE IMAGE OF $B \subseteq Y$

COMPOSITION OF FUNCTIONS

Consider the schematic representation of the functions $f: X \to Y$ and $g: Y \to Z$ in Figure 1.1, along with a third function $g \circ f: X \to T$.

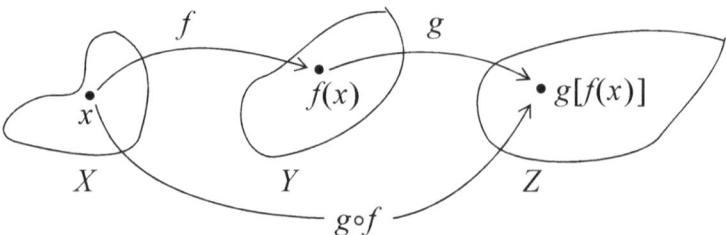

Figure 1.1

As is suggested in the above figure, the function $g \circ f: X \to Z$ is given by:
$$(g \circ f)(x) = g[f(x)]$$
↑ ↑
first apply f
and then apply g

1.1 Functions

Formally:

DEFINITION 1.4
COMPOSITION

Let $f: X \to Y$ and $g: Y \to Z$ be such that the range of f is contained in the domain of g. The composite function $g \circ f: X \to Z$ is given by:
$$(g \circ f)(x) = g[f(x)]$$

> Throughout the text the symbol $\Re$ will be used to denote the set of real numbers.

EXAMPLE 1.1 Let $f: \Re \to \Re$ and $g: \Re \to \Re$ be given by $f(x) = x^2 + 1$ and $g(x) = 2x - 5$. Find:

(a) $(g \circ f)(3)$ (b) $(f \circ g)(x)$

SOLUTION:

(a) $(g \circ f)(3) = g[f(3)] = g(3^2 + 1) = g(10) = 2 \cdot 10 - 5 = 15$

(b) $(f \circ g)(x) = f[g(x)] = f(2x - 5) = (2x - 5)^2 + 1 = 4x^2 - 20x + 26$

> $M_{2 \times 2} = \left\{ \begin{bmatrix} a & b \\ c & d \end{bmatrix} \Big| a, b, c, d \in \Re \right\}$
> (the set of two-by-two matrices)
>
> $\Re^2 = \{(a, b) | a, b \in \Re\}$
> (The set of two-tuples)
> In general:
> $\Re^n = \{(x_1, x_2, \ldots, x_n)\}$
> (The set of n-tuples)

EXAMPLE 1.2 let $f: M_{2 \times 2} \to \Re^2$ and $g: \Re^2 \to \Re$ (see margin) be given by $f\left(\begin{bmatrix} a & b \\ c & d \end{bmatrix} \right) = (ab, cd)$ and $g(a, b) = a - b$. Find:

(a) $(g \circ f)\left(\begin{bmatrix} 1 & 3 \\ 2 & 4 \end{bmatrix} \right)$ (b) $(g \circ f)\left(\begin{bmatrix} a & b \\ c & d \end{bmatrix} \right)$

SOLUTION:

(a) $(g \circ f)\left(\begin{bmatrix} 1 & 3 \\ 2 & 4 \end{bmatrix} \right) = g\left[f\left(\begin{bmatrix} 1 & 3 \\ 2 & 4 \end{bmatrix} \right) \right] = g(1 \cdot 3, 2 \cdot 4)$
$= g(3, 8) = 3 - 8 = -5$

(b) $(g \circ f)\left(\begin{bmatrix} a & b \\ c & d \end{bmatrix} \right) = g\left[f\left(\begin{bmatrix} a & b \\ c & d \end{bmatrix} \right) \right] = g(ab, cd) = ab - cd$

CHECK YOUR UNDERSTANDING 1.1

Let $f: M_{2 \times 2} \to \Re$ be given by $f\left(\begin{bmatrix} a & b \\ c & d \end{bmatrix} \right) = a + d$ and $g: \Re \to R^2$ be given by $g(x) = (2x, x^2)$. Determine:

(a) $(g \circ f)\left(\begin{bmatrix} 1 & 3 \\ 2 & 4 \end{bmatrix} \right)$ (b) $(g \circ f)\left(\begin{bmatrix} a & b \\ c & d \end{bmatrix} \right)$

Answer: (a) $(10, 25)$
(b) $(2a + 2d, a^2 + 2ad + d^2)$

BIJECTIONS AND THEIR INVERSES

DEFINITION 1.5   A function $f: X \to Y$ is:

ONE-TO-ONE   **One-to-one** if $f(a) = f(b) \Rightarrow a = b$

ONTO   **Onto** if for every $y \in Y$ there exists $x \in X$ such that $f(x) = y$.

BIJECTION   A **bijection** if it is both one-to-one and onto.

EXAMPLE 1.3   Let $f: \Re^4 \to M_{2 \times 2}$ be given by:

$$f(x, y, z, w) = \begin{bmatrix} -y & 2x \\ 3w & z \end{bmatrix}$$

Show that f is a bijection

SOLUTION: To show that f is one to one, we start with

$$f(x, y, z, w) = f(\bar{x}, \bar{y}, \bar{z}, \bar{w})$$

and go on to show that $(x, y, z, w) = (\bar{x}, \bar{y}, \bar{z}, \bar{w})$:

$$f(x,y,z,w) = f(\bar{x},\bar{y},\bar{z},\bar{w}) \Rightarrow \begin{bmatrix} -y & 2x \\ 3w & c \end{bmatrix} = \begin{bmatrix} -\bar{y} & 2\bar{x} \\ 3\bar{w} & c \end{bmatrix} \Rightarrow \left.\begin{matrix} -y = -\bar{y} \\ 2x = 2\bar{x} \\ 3w = 3\bar{w} \\ z = \bar{z} \end{matrix}\right\} \Rightarrow \left.\begin{matrix} y = \bar{y} \\ x = \bar{x} \\ w = \bar{w} \\ z = \bar{z} \end{matrix}\right\}$$

$$\Rightarrow (x, y, z, w) = (\bar{x}, \bar{y}, \bar{z}, \bar{w})$$

To show that f is onto, we take an arbitrary element $\begin{bmatrix} a & b \\ c & d \end{bmatrix} \in M_{2 \times 2}$ and set our sights on finding $(x, y, z, w) \in R^4$ such that $f(x, y, z, w) = \begin{bmatrix} a & b \\ c & d \end{bmatrix}$:

$$f(x,y,z,w) = \begin{bmatrix} a & b \\ c & d \end{bmatrix} \Rightarrow \begin{bmatrix} -y & 2x \\ 3w & z \end{bmatrix} = \begin{bmatrix} a & b \\ c & d \end{bmatrix} \Rightarrow \left.\begin{matrix} -y = a \\ 2x = b \\ 3w = c \\ z = d \end{matrix}\right\} \Rightarrow \left.\begin{matrix} y = -a \\ x = b/2 \\ w = c/3 \\ z = d \end{matrix}\right\}$$

The above argument shows that f will map the element $\left(\dfrac{b}{2}, -a, d, \dfrac{c}{3}\right) \in R^4$ to $\begin{bmatrix} a & b \\ c & d \end{bmatrix} \in M_{2 \times 2}$. Let's check it out:

$$f\left(\frac{b}{2}, -a, d, \frac{c}{3}\right) = \begin{bmatrix} -(-a) & 2\left(\frac{b}{2}\right) \\ 3\left(\frac{c}{3}\right) & d \end{bmatrix} = \begin{bmatrix} a & b \\ c & d \end{bmatrix}$$

CHECK YOUR UNDERSTANDING 1.2

(a) Show that the function $f: M_{2 \times 2} \to R^4$ given by
$f\left(\begin{bmatrix} a & b \\ c & d \end{bmatrix}\right) = (d, -c, 3a, b)$ is one-to-one and onto.

(b) Show that the function $f: M_{2 \times 2} \to M_{2 \times 2}$ given by
$f\left(\begin{bmatrix} a & b \\ c & d \end{bmatrix}\right) = \begin{bmatrix} b & a \\ c+d & 2b \end{bmatrix}$ is neither one-to-one nor onto.

Answer: See page A-1.

Consider the bijection $f: \{0, 1, 2, 3\} \to \{a, b, c, d\}$ depicted in Figure 1.2(a) and the function $f^{-1}: \{a, b, c, d\} \to \{0, 1, 2, 3\}$ in Figure 1.2(b). The function f^{-1}, called the **inverse of the function** f, was obtained from f by "reversing" the direction of the arrows in Figure 1.2(a).

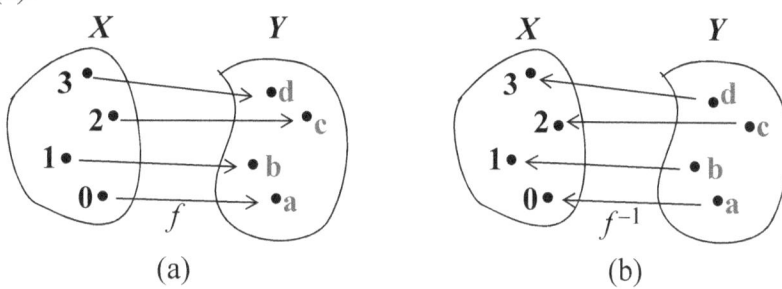

Figure 1.2

In general:

DEFINITION 1.6
INVERSE FUNCTION

The **inverse** of a bijection $f: X \to Y$, is the function $f^{-1}: Y \to X$ given by:
$$f^{-1}(y) = x \text{ where } f(x) = y$$
More formally:
$$f^{-1} = \{(y, x) \mid (x, y) \in f\}$$

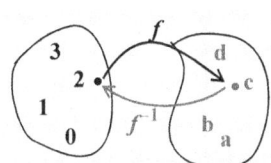

Returning to Figure 1.2, we observe that the inverse of the bijection f is also a bijection. We also note that if we apply f and then f^{-1} we will end up where we started, and ditto if we first apply f^{-1} and then f (see margin). In general:

THEOREM 1.1 Let $f: X \to Y$ be a bijection. Then:

(a) $f^{-1}: Y \to X$ is also a bijection.

(b) $f^{-1}[f(x)] = x \; \forall x \in X$ and
$f[f^{-1}(y)] = y \; \forall y \in Y$

PROOF: (a) f^{-1} **is one-to-one:** If $f^{-1}(y_1) = f^{-1}(y_2) = x$, then:

$$(y_1, x) \in f^{-1} \text{ and } (y_2, x) \in f^{-1}$$
$$\Rightarrow (x, y_1) \in f \text{ and } (x, y_2) \in f$$
$$\Rightarrow y_1 = y_2 \text{ (since } f \text{ is a function)}$$

f^{-1} **is onto:** Let $x \in X$. Since f is onto, there exists $y \in Y$ such that $(x, y) \in f$. Then: $(y, x) \in f^{-1} \Rightarrow f^{-1}(y) = x$.

(b) Let $x \in X$. Since $[x, f(x)] \in f$, $[f(x), x] \in f^{-1}$, which is to say: $x = f^{-1}[f(x)]$. As for the other direction:

Recall that to say that $f(x) = y$ is to say that $(x, y) \in f$. (see Definition 1.3).

CHECK YOUR UNDERSTANDING 1.3

Verify that for any bijection $f: X \to Y$:

$$f[f^{-1}(y)] = y \ \forall y \in Y$$

Answer: See page A-2.

EXAMPLE 1.4 (a) Find the inverse of the binary function
$f: \mathfrak{R}^4 \to M_{2 \times 2}$ given by:

$$f(x, y, z, w) = \begin{bmatrix} -y & 2x \\ 3w & z \end{bmatrix}$$

(see Example 1.3)

(b) Show, directly, that

$$f\left[f^{-1}\left(\begin{bmatrix} a & b \\ c & d \end{bmatrix}\right)\right] = \begin{bmatrix} a & b \\ c & d \end{bmatrix}$$

SOLUTION: (a) For given $\begin{bmatrix} a & b \\ c & d \end{bmatrix}$ we determine (x, y, z, w) such that $f(x, y, z, w) = \begin{bmatrix} a & b \\ c & d \end{bmatrix}$:

$$f(x, y, z, w) = \begin{bmatrix} a & b \\ c & d \end{bmatrix} \Rightarrow \begin{bmatrix} -y & 2x \\ 3w & z \end{bmatrix} = \begin{bmatrix} a & b \\ c & d \end{bmatrix} \Rightarrow \left.\begin{array}{l} -y = a \\ 2x = b \\ 3w = c \\ z = d \end{array}\right\} \Rightarrow \left.\begin{array}{l} y = -a \\ x = b/2 \\ w = c/3 \\ z = d \end{array}\right\}$$

Conclusion: $f^{-1}\left(\begin{bmatrix} a & b \\ c & d \end{bmatrix}\right) = \left(\frac{b}{2}, -a, d, \frac{c}{3}\right)$

> Note: In Example 1.3 we showed that for $f(x,y,z,w) = \begin{bmatrix} -y & 2x \\ 3w & z \end{bmatrix}$:
>
> $$f\left(\frac{b}{2}, -a, d, \frac{c}{3}\right) = \begin{bmatrix} a & b \\ c & d \end{bmatrix}$$
>
> That being the case: $f^{-1}\left(\begin{bmatrix} a & b \\ c & d \end{bmatrix}\right) = \left(\frac{b}{2}, -a, d, \frac{c}{3}\right)$.

(b) $f\left[f^{-1}\left(\begin{bmatrix} a & b \\ c & d \end{bmatrix}\right)\right] = f\left(\frac{b}{2}, -a, d, \frac{c}{3}\right) = \begin{bmatrix} -(-a) & 2\left(\frac{b}{2}\right) \\ 3\left(\frac{c}{3}\right) & d \end{bmatrix} = \begin{bmatrix} a & b \\ c & d \end{bmatrix}$

since $f(x,y,z,w) = \begin{bmatrix} -y & 2x \\ 3w & z \end{bmatrix}$

CHECK YOUR UNDERSTANDING 1.4

Find the inverse of the bijection $f: M_{2 \times 2} \to R^4$ given by $f\left(\begin{bmatrix} a & b \\ c & d \end{bmatrix}\right) = (d, -c, 3a, b)$ and verify, directly, that:

$$f[f^{-1}(x,y,z,w)] = (x,y,z,w) \text{ and that } f^{-1}\left[f\begin{bmatrix} a & b \\ c & d \end{bmatrix}\right] = \begin{bmatrix} a & b \\ c & d \end{bmatrix}.$$

Answer:
$f^{-1}(x,y,z,w) = \begin{bmatrix} z/3 & w \\ -y & x \end{bmatrix}$
For the rest: See page A-2.

As it turns out, one-to-one and onto properties are preserved under composition:

THEOREM 1.2 Let $f: X \to Y$ and $g: Y \to Z$ be functions with the range of f contained in the domain of g. Then:

(a) If f and g are one-to-one, so is $g \circ f$.

(b) If f and g are onto, so is $g \circ f$.

(c) If f and g are bijections, so is $g \circ f$.

PROOF: (a) Assume that both f and g are one-to-one, and that:

$$(g \circ f)(x_1) = (g \circ f)(x_2)$$

Which is to say: $g[f(x_1)] = g[f(x_2)]$

Since g is one-to-one: $f(x_1) = f(x_2)$

Since f is one-to-one: $x_1 = x_2$

(b) Assume that both *f* and *g* are onto, and let $z \in Z$. We are to find $x \in X$ such that $(g \circ f)(x) = z$. Let's do it:

Since *g* is onto, there exists $y \in Y$ such that $g(y) = z$.
Since *f* is onto, there exists $x \in X$ such that $f(x) = y$.
It follows that $(g \circ f)(x) = g[f(x)] = g(y) = z$.

(c) If *f* and *g* are both bijections then, by (a) and (b), so is $g \circ f$.

Theorem 1.2(c) asserts that the composition $g \circ f$ of two bijections is again a bijection. As such, it has an inverse, and here is how it is related to the inverses of its components:

THEOREM 1.3 If $f: X \to Y$ and $g: Y \to Z$ are bijections, then:
$$(g \circ f)^{-1} = f^{-1} \circ g^{-1}$$

This is an example of a so-called "shoe-sock theorem." Why the funny name?

One puts on socks then shoes in the reverse process: The shoes come off and then the socks.

PROOF: For given $z \in Z$, let $x \in X$ be such that $(g \circ f)(x) = z$; which is to say, that $(g \circ f)^{-1}(z) = x$. We complete the proof by showing that $(f^{-1} \circ g^{-1})(z)$ is also equal to *x*:

$$(g \circ f)^{-1}(z) = x$$
$$z = (g \circ f)(x)$$
$$z = g[f(x)]$$
$$g^{-1}(z) = f(x)$$
$$f^{-1}[g^{-1}(z)] = x$$
$$(f^{-1} \circ g^{-1})(z) = x$$

CHECK YOUR UNDERSTANDING 1.5

The function $f: \Re^4 \to M_{2 \times 2}$ given by $f(x, y, z, w) = \begin{bmatrix} -y & 2x \\ 3w & z \end{bmatrix}$ has inverse $f^{-1}\left(\begin{bmatrix} a & b \\ c & d \end{bmatrix}\right) = \left(\frac{b}{2}, -a, d, \frac{c}{3}\right)$ (see Example 1.4), and the function $g\left(\begin{bmatrix} a & b \\ c & d \end{bmatrix}\right) = (d, -c, 3a, b)$ has inverse $g^{-1}(x, y, z, w) = \begin{bmatrix} z/3 & w \\ -y & x \end{bmatrix}$.

Determine the function $g \circ f: \Re^4 \to \Re^4$ and its inverse; and then show, directly, that $(g \circ f)^{-1} = f^{-1} \circ g^{-1}$.

Answer: See page A-2.

EXERCISES

Exercises 1-19. Let $U = \{1, 2, 3, ...\}$ $O = \{1, 3, 5, ...\}$ $E = \{2, 4, 6, ...\}$,
$A = \{5n | n \in U\}$ $B = \{3n | n \in U\}$ $C = \{1, 2, 3, ..., 15\}$,
$D = \{2, 4, 6, ... 10\}$ $F = \{11, 12, 13, 14\}$. Determine:

1. $O \cup E$
2. $O \cap E$
3. $A \cap B$
4. $A \cup B$
5. $B \cup C$
6. $B \cap C$
7. $C \cup D$
8. $C \cap D$
9. $O^c \cup E^c$
10. $O^c \cap A$
11. $C \cap O$
12. $(O \cap A)^c$
13. $(C \cap D) \cup F$
14. $C \cap (D \cup F)$
15. $(C \cup F) \cap D$
16. $(C \cap F^c) \cup F$
17. $(B^c \cap C) \cup (D \cap O)$
18. $[(O \cup E)^c \cup (A \cap B)]^c$
19. $[(O \cap E)^c \cap (O \cup A)]^c$

20. Establish the following set identities (all capital letters represent subsets of a universal set U):

(a) DeMorgan's Theorems:

(i) $(A \cap B)^c = A^c \cup B^c$ (ii) $(A \cup B)^c = A^c \cap B^c$

(b) Associative Theorems:

(i) $A \cup (B \cup C) = (A \cup B) \cup C$ (ii) $A \cap (B \cap C) = (A \cap B) \cap C$

(c) Distributive Theorems:

(i) $A \cap (B \cup C) = (A \cap B) \cup (A \cap C)$ (ii) $A \cup (B \cap C) = (A \cup B) \cap (A \cup C)$

Exercises 21-23. Prove that:

21. $[A^c \cup B]^c = A \cap B^c$
22. $(A \cap B^c)^c \cup B = A^c \cup B$
23. $(A \cap B) \cup (A \cap B^c) = A$

Exercises 24-26. Give a counterexample to show that each of the following statements is **False**.

24. $(A \cap B)^c = A^c \cap B^c$
25. $(A \cup B)^c = A^c \cup B^c$
26. $(A \cap B)^c \cap C^c = A^c \cap (B \cap C)^c$

Exercises 27-30. Is $f: \Re \to \Re$ (a) One-to-one? (b) Onto?

27. $f(x) = \dfrac{3x - 7}{x + 2}$
28. $f(x) = x^2 - 3$
29. $f(x) = \dfrac{x^2 + 1}{x^4 + 1}$
30. $f(x) = x^3 - x + 2$

Exercises 31-33. Is $f: \Re \to \Re^2$ (a) One-to-one? (b) Onto?

31. $f(x) = (x, x)$
32. $f(x) = (x, 1)$
33. $f(x) = (x^2 + 2x, x + 5)$

Exercises 34-36. Is $f: \Re^2 \to \Re^2$ (a) One-to-one? (b) Onto?

34. $f(x, y) = (y, -x)$
35. $f(x, y) = (x, x + y)$
36. $f(x, y) = (2x, x + y)$

Exercises 37-38. Is $f\colon M_{2\times 2} \to \Re^4$ (a) One-to-one? (b) Onto?

37. $f\left(\begin{bmatrix} a & b \\ c & d \end{bmatrix}\right) = (a, -2b, c, c-d)$

38. $f\left(\begin{bmatrix} a & b \\ c & d \end{bmatrix}\right) = (a-b, c, d, b-a)$

Exercises 39-40. Is $f\colon \Re^4 \to M_{2\times 2}$ (a) One-to-one? (b) Onto?

39. $f(a, b, c, d) = \begin{bmatrix} ab & b+a \\ c+b & a^2 b^2 \end{bmatrix}$

40. $f(a, b, c, d) = \begin{bmatrix} a & b+a \\ c+b & d+a \end{bmatrix}$

Exercises 41-49. Show that the given function $f\colon X \to Y$ is a bijection. Determine $f^{-1}\colon Y \to X$ and show, directly, that $(f^{-1}\circ f)(x) = x$ $\forall x \in X$ and that $(f\circ f^{-1})(y) = y$ $\forall y \in Y$.

41. $X = \Re$, $Y = \Re$, and $f(x) = 3x - 2$.

42. $X = (-\infty, 0) \cup (0, \infty)$, $Y = (-\infty, 1) \cup (1, \infty)$, and $f(x) = \dfrac{x+1}{x}$.

43. $X = (-\infty, -1) \cup (-1, \infty)$, $Y = (-\infty, 2) \cup (2, \infty)$, and $f(x) = \dfrac{2x}{x+1}$.

44. $X = Y = \Re^2$, and $f(a, b) = (-b, a)$.

45. $X = Y = \Re^2$, and $f(a, b) = (5a, b+3)$.

46. $X = Y = M_{2\times 2}$, and $f\left(\begin{bmatrix} a & b \\ c & d \end{bmatrix}\right) = \begin{bmatrix} b & c \\ d & a \end{bmatrix}$.

47. $X = Y = M_{2\times 2}$, and $f\left(\begin{bmatrix} a & b \\ c & d \end{bmatrix}\right) = \begin{bmatrix} c & 2d \\ a & -b \end{bmatrix}$.

48. $X = \Re^4$, $Y = M_{2\times 2}$, and $f(a, b, c, d) = \begin{bmatrix} 2b & c+1 \\ d & -a \end{bmatrix}$.

49. $X = M_{3\times 1}$, $Y = \Re^3$, and $f\left(\begin{bmatrix} a \\ b \\ c \end{bmatrix}\right) = (2a, a-b, b+c)$.

50. Prove that a function $f\colon \Re \to \Re$ is one-to-one if and only if the function $g\colon \Re \to \Re$ given by $g(x) = -f(x)$ is one-to-one.

51. Prove that for any given $f\colon X \to Y$, $g\colon Y \to S$, and $h\colon S \to T$: $h\circ(g\circ f) = (h\circ g)\circ f$.

52. Let $f: X \to Y$, $g: X \to Y$, and $h: Y \to W$ be given, with h a bijection.
    (a) Prove that if $h \circ f = h \circ g$, then $f = g$.
    (b) Show, by means of an example, that (a) need not hold when h is not a bijection.

53. Let $S \subseteq X$, $Y \neq \emptyset$, and $f: S \to Y$ be given. Prove that there exists a function $g: X \to Y$ such that $f(x) = g(x)$ for every $x \in S$. (That is, a function g which "extends" f to all of X.)

54. Let $S \subseteq X$, $Y \neq \emptyset$, and $f: X \to Y$ be given. Prove that there exists a function $g: S \to Y$ such that $f(x) = g(x)$ for every $x \in S$. (That is, a function g which is the "restriction" of f to the subset S.)

Exercise. 55-60. (Algebra of Functions) For any set X, and functions $f: X \to \Re$ and $g: X \to \Re$, we define $f+g: X \to \Re$, $f-g: X \to \Re$, $f \cdot g: X \to \Re$, and $\frac{f}{g}: X \to \Re$ as follows:

$$(f+g)(x) = f(x) + g(x) \qquad (f-g)(x) = f(x) - g(x)$$
$$(f \cdot g)(x) = f(x) \cdot f(x) \qquad \left(\frac{f}{g}\right)(x) = \frac{f(x)}{g(x)} \text{ if } g(x) \neq 0$$

55. Prove that for any $f: X \to \Re$ and $g: X \to \Re$: $f+g = g+f$ and $f \cdot g = g \cdot f$.
56. Exhibit $f: \Re \to \Re$, $g: \Re \to \Re$, such that $f-g \neq g-f$.
57. Exhibit one-to-one functions $f: \Re \to \Re$, $g: \Re \to \Re$, such that $f+g$ is not one-to-one.
58. Exhibit onto functions $f: \Re \to \Re$, $g: \Re \to \Re$, such that $f+g$ is not onto.
59. Exhibit one-to-one functions $f: \Re \to \Re$, $g: \Re \to \Re$, such that $f \cdot g$ is not one-to-one.
60. Exhibit onto functions $f: \Re \to \Re$, $g: \Re \to \Re$, such that $f+g$ is not onto.

	PROVE OR GIVE A COUNTEREXAMPLE	

61. If $A \cap B \neq \emptyset$ and $B \cap C \neq \emptyset$, then $A \cap C \neq \emptyset$.
62. If $A \cap B = \emptyset$ or $B \cap C = \emptyset$, then $A \cap C = \emptyset$.
63. If $A \subseteq (B \cap C)$ and $C \subseteq (B \cap A)$, then $A = C$.
64. If $A \cup B = A \cup C$, then $A = C$.
65. If $A \cap B = A \cap C$, then $A = C$.
66. If $A \cap B = A \cup B$, then either $A = \emptyset$ or $B = \emptyset$.
67. If $A \subseteq (B \cap C)$ and $B \subseteq (C \cap D)$, then $(A \cap C) \subseteq (B \cap D)$.

68. $(A \cup B) \cap (A \cup C) = A \cup (B \cap C)$.

69. If no element of a set A is contained in a set B, then A cannot be a subset of B.

70. Two sets A and B are equal if and only if the set of all subsets of A is equal to the set of all subsets of B.

71. $\varnothing = \{\varnothing\}$.

§2. Principle of Mathematical Induction

This section introduces a most powerful mathematical tool, the Principle of Mathematical Induction (*PMI*). Here is how it works:

A form of the Principle of Mathematical Induction is actually one of Peano's axioms, which serve to define the positive integers.
[Giuseppe Peano (1858-1932).]

PMI	
Let $P(n)$ denote a proposition that is either true or false, depending on the value of the integer n.	
If:	I. $P(1)$ is True.
And if, from the **assumption** that:	II. $P(k)$ is True
one can show that:	III. $P(k+1)$ is also True.
then the proposition $P(n)$ is valid **for all integers** $n \geq 1$	

Step II of the induction procedure may strike you as being a bit strange. After all, if one can assume that the proposition is valid at $n = k$, why not just assume that it is valid at $n = k+1$ and save a step! Well, you can assume whatever you want in Step II, but if the proposition is not valid for all n you simply are not going to be able to demonstrate, in Step III, that the proposition holds at the next value of n. Just imagine that the propositions

$$P(1), P(2), P(3), \ldots, P(k), P(k+1), \ldots$$

are lined up, as if they were an infinite set of dominoes:

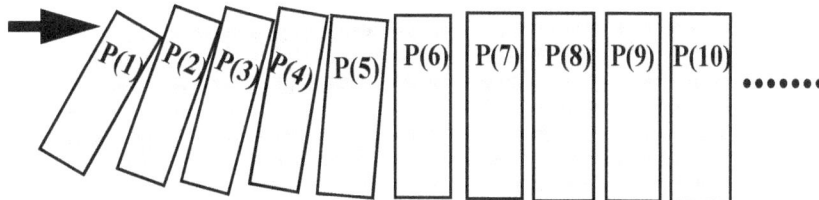

If you knock over the first domino (Step I), and if when a domino falls (Step II) it knocks down the next one (Step III), then all of the dominoes will surely fall. But if the falling k^{th} domino fails to knock over the next one, then all the dominoes need not fall.

The *Principle of Mathematical Induction* might have been better labeled the *Principle of Mathematical Deduction*, for inductive reasoning is used to formulate a hypothesis or conjecture, while deductive reasoning is used to rigorously establish whether or not the conjecture is valid.

To illustrate how the process works, we ask you to consider the sum of the first n odd integers, for $n = 1$ through $n = 5$:

Sum of the first n odd integers	Sum
1	1
1 + 3	4
1 + 3 + 5	9
1 + 3 + 5 + 7	16
1 + 3 + 5 + 7 + 9	25

n	Sum
1	1
2	4
3	9
4	16
5	25
6	?

Figure 1.3

Looking at the pattern of the table on the right in Figure 1.3, you can probably anticipate that the sum of the first 6 odd integers will turn out to be $6^2 = 36$, which is indeed the case. Indeed, the pattern suggests that: The sum of the first n odd integers is n^2

Using the Principle of Mathematical Induction, we now establish the validity of the above conjecture:

Let $P(n)$ be the proposition that the sum of the first n odd integers equals n^2.

I. Since the sum of the first 1 odd integers is 1^2, $P(1)$ is true.

II. **Assume** $P(k)$ is true; that is:

$$1 + 3 + 5 + \cdots + (2k - 1) = k^2$$

see margin ↗

III. We show that $P(k + 1)$ is true, thereby completing the proof:

the sum of the first $k + 1$ odd integers

$$[1 + 3 + 5 + \cdots + (2k - 1)] + (2k + 1) = k^2 + (2k + 1) = (k + 1)^2$$

induction hypothesis: Step II

margin:

The sum of the first **3** odd integers is:

$1 + 3 + 5 \leftarrow \boxed{2 \cdot 3 - 1}$

The sum of the first **4** odd integers is:

$1 + 3 + 5 + 7 \leftarrow \boxed{2 \cdot 4 - 1}$

Suggesting that the sum of the first k odd integers is:

$1 + 3 + \ldots + \boxed{(2k - 1)}$

(see Exercise 1).

EXAMPLE 1.5 Use the Principle of Mathematical Induction to establish the following formula for the sum of the first n integers:

$$1 + 2 + 3 + \ldots + n = \frac{n(n + 1)}{2}$$

SOLUTION: Let $P(n)$ be the proposition:

$$1 + 2 + 3 + \ldots + n = \frac{n(n + 1)}{2} \quad (*)$$

I. $P(1)$ is true: $1 = \frac{1(1 + 1)}{2}$ Check!

II. **Assume** $P(k)$ is true: $\mathbf{1 + 2 + 3 + \ldots + k} = \dfrac{\mathbf{k(k + 1)}}{\mathbf{2}}$

III. We are to show that $P(k + 1)$ is true; which is to say, that (*) holds when $n = k + 1$:

$$1 + 2 + 3 + \ldots + k + (k + 1) = \frac{(k + 1)[(k + 1) + 1]}{2} = \frac{(k + 1)(k + 2)}{2}$$

Let's do it:

$$1 + 2 + 3 + \ldots + k + (k + 1) = [1 + 2 + 3 + \cdots + k] + (k + 1)$$

induction hypothesis: $= \dfrac{k(k + 1)}{2} + (k + 1)$

$$= \frac{k(k + 1) + 2(k + 1)}{2} = \frac{(k + 1)(k + 2)}{2}$$

> **CHECK YOUR UNDERSTANDING 1.6**
>
> (a) Use the formula for the sum of the first n odd integers, along with that for the sum of the first n integers, to derive a formula for the sum of the first n even integers.
>
> (b) Use the Principle of Mathematical Induction directly to establish the formula you obtained in (a).

Answer: See page A-3.

We pause momentarily to recall three number theory definitions. In the present discussion, Z denotes the set of integers.

DEFINITION 1.7
EVEN AND ODD

$n \in Z$ is **even** if $\exists k \in Z \ni n = 2k$.

$n \in Z$ is **odd** if $\exists k \in Z \ni n = 2k+1$.

DIVISIBILITY

A nonzero integer a **divides** $b \in Z$, written $a|b$, if $b = ak$ for some $k \in Z$.

THEOREM 1.4 Let b and c be nonzero integers. Then:

(a) If $a|b$ and $b|c$, then $a|c$.

(b) If $a|b$ and $a|c$, then $a|(b+c)$.

(c) If $a|b$, then $a|bc$ for every c.

PROOF: (a) If $a|b$ and $b|c$, then, by Definition 1.7:
$$b = ak \text{ and } c = bh \text{ for some } h \text{ and } k.$$
Consequently:
$$c = bh = (ak)h = a(kh) = at \text{ (where } t = kh\text{)}.$$
It follow, from Definition 1.7, that $a|c$.

> Note how Definition 1.7 is used in both directions in the above proof.

(b) If $a|b$ and $a|c$, then $b = ah$ and $c = ak$ for some h and k. Consequently:
$$b+c = ah + ak = a(h+k) = at \text{ (where } t = h+k\text{)}.$$
It follows that $a|(b+c)$.

(c) If $a|b$, then $b = ak$ for some k. Consequently, for any c:
$$bc = (ak)c = a(kc) = at \text{ (where } t = kc\text{)}.$$
It follows that $a|bc$.

> **CHECK YOUR UNDERSTANDING 1.7**
>
> Prove or give a counterexample.
>
> (a) If $a|(b+c)$, then $a|b$ or $a|c$.
>
> (b) If $a|b$ and $a|(b+c)$, then $a|c$.

Answer: See page A-3.

The "domino effect" of the Principle of Mathematical Induction need not start by knocking down the first domino $P(1)$. Consider the following example where domino $P(0)$ is the first to fall.

EXAMPLE 1.6 Use the Principle of Mathematical Induction to show that $4|(5^n - 1)$ for all integers $n \geq 0$.

SOLUTION: Let $P(n)$ be the proposition $4|(5^n - 1)$.

I. $P(0)$ is true: $4|(5^0 - 1)$, since $5^0 - 1 = 1 - 1 = 0$.

II. **Assume** $P(k)$ is true: $4|(5^k - 1)$.

III. We show $P(k+1)$ is true; namely, that $4|(5^{k+1} - 1)$:

$$5^{k+1} - 1 = 5(5^k) - \mathbf{1} = 5(5^k) - \mathbf{5 + 4} \quad \text{(see margin)}$$

$$= 5(5^k - 1) + 4$$

The desired conclusion now follows from Theorem 1.4:

Theorem 1.4 (c): $4|(5^k - 1) \Rightarrow 4|5(5^k - 1)$ and then:

Theorem 1.4(b): $4|5(5^k - 1)$ and $4|4 \Rightarrow 4|[5(5^k - 1) + 4]$

What motivated us to write -1 *in the form* $-5 + 4$ *? Necessity did:*

We had to do something to get "$5^k - 1$" into the picture (see II).

Clever, to be sure; but such a clever move stems from stubbornly focusing on what is given and on what needs to be established.

CHECK YOUR UNDERSTANDING 1.8

(a) Use the Principle of Mathematical Induction to show that $n! > n^2$ for all integers $n \geq 4$.

(b) Use the Principle of Mathematical Induction to show that $6|(n^3 + 5n)$ for all integers $n \geq 1$.

Answer: See page A-4.

ALTERNATE FORMS OF MATHEMATICAL INDUCTION

We complete this section by introducing two equivalent forms of the Principle of Mathematical Induction — equivalent in that any one of them can be used to establish the remaining two.

One version, which we will call the Alternate Principle of Induction (API), is displayed in Figure 1.3(b). As you can see, the only difference between PMI and API surfaces in (*) and (**). Specifically, the proposition "$P(k)$ True" in (a) is replaced, in (b), with the proposition "$P(m)$ True for all integers m **up to and including** k".

API is often called the Strong Principle of Induction. A bit of a misnomer, since it is, in fact, equivalent to PMI.

Let $P(n)$ denote a proposition that is either true or false, depending on the value of the integer n.	
PMI	**API**
If $P(1)$ is True, and if:	If $P(1)$ is True, and if
(*) $P(k)$ **True** $\Rightarrow P(k+1)$ **True**	(**): $P(m)$ **True** for $1 \leq m \leq k \Rightarrow P(k+1)$ **True**
then $P(n)$ is True for all integers $n \geq 1$	then $P(n)$ is True for all integers $n \geq 1$
(a)	(b)

Figure 1.4

We establish the equivalence of PMI and API by showing that (*) holds if and only (**) holds. Clearly, if (*) holds then (**) must also hold. As for the other way around:

Assume that (**) holds and that (*) does not.
(we will arrive at a contradiction)

If (*) does not hold, then there must exist some k_0 for which $P(k_0)$ is True and $P(k_0 + 1)$ is False. Since $P(k_0 + 1)$ is False, and since (**) holds, we know that $P(k_1)$ is False for some $1 \leq k_1 \leq k_0$. But we are assuming that $P(k_0)$ is True. Hence $P(k_1)$ is False for some $1 \leq k_1 < k_0$.

Repeating the above procedure with k_1 playing the role of k_0 we arrive at $P(k_2)$ is False for some $1 \leq k_2 < k_1$.

Continuing in this fashion we shall, after at most $k_0 - 1$ steps, be forced to conclude that $P(1)$ is False — contradicting the assumption that $P(1)$ is True.

EXAMPLE 1.7 Use API to show that for any given integer $n \geq 12$ there exist integers $a > 0, b \geq 0$ such that
$$n = 3a + 7b.$$

SOLUTION:

I. Claim holds for $n = 12$: $12 = 3 \cdot 4 + 7 \cdot 0$

II. Assume claim holds for all m such that $12 \leq m \leq k$.

III. To show that the claim holds for $n = k+1$ we first show, directly, that it does indeed hold if $k+1 = 13$ or if $k+1 = 14$:
$$13 = 3 \cdot 2 + 7 \cdot 1 \text{ and } 14 = 3 \cdot 0 + 7 \cdot 2$$
Now consider any $k + 1 \geq 15$.
If $k + 1 \geq 15$, then $12 \leq (k+1) - 3 \leq k$. Appealing to the induction hypothesis, we choose $a > 0, b \geq 0$ such that:
$$(k+1) - 3 = 3a + 7b$$
It follows that $k + 1 = 3(a+1) + 7b$, and the proof is complete.

Here is another important property which turns out to be equivalent to the Principle of Mathematical Induction:

> Z^+ denotes the set of positive integers.

> Note that subsets of Z need not have first elements. A case in point
> $\{..., -4, -2, 0, 2, 4, ...\}$
> Note also that the bounded set
> $\{x \in \Re | 5 < x < 9\}$
> does not contain a smallest element (5 is not in the set).

THE WELL-ORDERING PRINCIPLE FOR Z^+

Every nonempty subset of Z^+ has a smallest (or least, or first) element.

We show that the Alternate Principle of Mathematical Induction implies the Well-Ordering Principle:

Let S be a NONEMPTY subset of Z^+.

If $1 \in S$, then it is certainly the smallest element in S, and we are done.

Assume $1 \notin S$, and suppose that S **does not** have a smallest element (we will arrive at a contradiction):

Let $P(n)$ be the proposition that $n \notin S$ for $n \in Z^+$. Since, $1 \notin S$, $P(1)$ is True. Suppose that $P(m)$ is True for all $1 \leq m \leq k$, can $P(k+1)$ be False? No:

To say that $P(k+1)$ is False is to say that $k+1 \in S$. But that would make $k+1$ the smallest element in S, since none of its predecessors are in S. This cannot be, since S was assumed not to have a smallest element.

Since $P(1)$ is True ($1 \notin S$) and since the validity of $P(m)$ for all $1 \leq m \leq k$ implies the validity of $P(k+1)$, $P(n)$ must be True for all $n \in Z^+$; which is the same as saying that no element of Z^+ is in S — contradicting the assumption that S is NONEMPTY.

CHECK YOUR UNDERSTANDING 1.9

Show that the Well-Ordering Principle implies the Principle of Mathematical Induction.

Answer: See page A-4.

1.2 Principle of Mathematical Induction 19

| | **EXERCISES** | |

Exercises 1-29. Establish the validity of the given statement.

1. For every integer $n \geq 1$, $2n - 1$ is the nth odd integer.

2. For every integer $n \geq 1$, $1 + 4 + 7 + \cdots + (3n - 2) = \dfrac{3n^2 - n}{2}$.

3. For every integer $n \geq 1$, $1^2 + 3^2 + 5^2 + \cdots + (2n - 1)^2 = \dfrac{n(2n - 1)(2n + 1)}{3}$.

4. For every integer $n \geq 1$, $1^2 + 2^2 + 3^2 + \cdots + n^2 = \dfrac{n(n + 1)(2n + 1)}{6}$.

5. For every integer $n \geq 1$, $4 + 4^2 + 4^3 + \cdots + 4^n = \dfrac{4(4^n - 1)}{3}$.

6. For every integer $n \geq 1$, $\dfrac{1}{2} + \dfrac{1}{4} + \dfrac{1}{8} + \cdots + \dfrac{1}{2^n} = 1 - \dfrac{1}{2^n}$.

7. For every integer $n \geq 1$, $\left(1 + \dfrac{1}{1}\right)\left(1 + \dfrac{1}{2}\right)\left(1 + \dfrac{1}{3}\right)\cdots\left(1 + \dfrac{1}{n}\right) = n + 1$.

8. For every integer $n \geq 1$ and any real number $x \neq 1$, $x^0 + x^1 + x^2 + \cdots + x^n = \dfrac{1 - x^{n+1}}{1 - x}$.

9. For every integer $n \geq 1$, and any real number $r \neq 1$, $\sum_{i=0}^{n} ar^i = \dfrac{a(1 - r^{n+1})}{1 - r}$.

10. For every integer $n \geq 0$: $5 | (2^{4n+2} + 1)$.

11. For every integer $n \geq 1$: $9 | (4^{3n} - 1)$.

12. For every integer $n \geq 1$: $3 | (5^n - 2^n)$.

13. For every integer $n \geq 1$, $5^{2n} + 7$ is divisible by 8.

14. For every integer $n \geq 1$, $3^{3n+1} + 2^{n+1}$ is divisible by 5.

15. For every integer $n \geq 1$, $4^{n+1} + 5^{2n-1}$ is divisible by 21.

16. For every integer $n \geq 1$, $3^{2n+2} - 8n - 9$ is divisible by 64.

17. For every integer $n \geq 0$, $2^n > n$.

18. For every integer $n \geq 5$, $2n - 4 > n$.

19. For every integer $n \geq 5$, $2^n > n^2$.

20. For every integer $n \geq 4$, $3^n > 2^n + 10$.

21. For every integer $n \geq 1$, $\dfrac{(2n)!}{2^n n!}$ is an odd integer.

22. For every integer $n \geq 4$, $2n < n!$.

23. (Calculus Dependent) Show that the sum of n differentiable functions is again differentiable.

24. (Calculus Dependent) Show that for every integer $n \geq 1$, $\dfrac{d}{dx} x^n = n x^{n-1}$.

    Suggestion: Use the product Theorem: If f and g are differentiable functions, then so is $f \cdot g$ differentiable, and $\dfrac{d}{dx}[f(x)g(x)] = f(x)\dfrac{d}{dx}g(x) + g(x)\dfrac{d}{dx}f(x)$.

25. Let $a_1 = 1$ and $a_{n+1} = 3 - \dfrac{1}{a_n}$. Show that $a_{n+1} > a_n$.

26. Let $a_1 = 2$ and $a_{n+1} = \dfrac{1}{3 - a_n}$. Show that $a_{n+1} < a_n$.

27. For every integer $n \geq 1$, $1 + \dfrac{1}{\sqrt{2}} + \dfrac{1}{\sqrt{3}} + \cdots + \dfrac{1}{\sqrt{n}} > 2(\sqrt{n+1} - 1)$.

28. For any positive number x, $(1 + x)^n \geq 1 + nx$ for every $n \geq 1$.

29. For every integer $n \geq 8$, there exist integers $a > 0$, $b > 0$ such that $n = 3a + 5b$.

30. Let m be any nonnegative integer. Use the Well-Ordering Principle to show that every nonempty subset of the set $\{n \in Z \mid n \geq -m\}$ contains a smallest element.

31. Use the Principle of Mathematical Induction to show that there are $n!$ different ways of ordering n objects, where $n! = 1 \cdot 2 \cdot 3 \cdot \ldots \cdot n$.

32. What is wrong with the following "Proof" that any two positive integers are equal:

    Let $P(n)$ be the proposition: *If a and b are any two positive integers such that* $\max(a, b) = n$, *then* $a = b$.

    I. $P(1)$ is true: If $\max(a, b) = 1$, then both a and b must equal 1.
    II. Assume $P(k)$ is true: If $\max(a, b) = k$, then $a = b$.
    III. We show $P(k + 1)$ is true:
       If $\max(a, b) = k + 1$ then $\max(a - 1, b - 1) = k$.
       By II, $a - 1 = b - 1 \Rightarrow a = b$.

§3. The Division Algorithm and Beyond

ALL LETTERS IN THIS SECTION WILL BE UNDERSTOOD TO REPRESENT INTEGERS.

In elementary school you learned how to divide one integer into another to arrive at a quotient and a remainder, and could then check your answer (see margin). That checking process reveals an important result:

THEOREM 1.5
THE DIVISION ALGORITHM

For any given $a \in Z$ and $d \in Z^+$, there exist unique integers q and r, with $0 \le r < d$, such that:
$$a = dq + r$$

Margin:

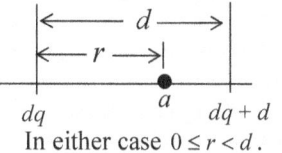

Check: $17 = 3 \cdot 5 + 2$
$a = dq + r$

Here is a "convincing argument" for your consideration:
Mark off multiples of d on the number line:

$\begin{array}{ccccc} | & | & | & | & | \\ -2d & -d & 0 & d & 2d \end{array}$

Case 1. If $a = dq$, then let $r = 0$.
Case 2. If a is not a multiple of d, then let dq be such that $dq < a < (d+1)q$. We then have $a = dq + r$, where:

$\begin{array}{c} \longleftarrow d \longrightarrow \\ \longleftarrow r \longrightarrow \bullet \\ dq \quad\quad a \quad dq+d \end{array}$

In either case $0 \le r < d$.

PROOF: We begin by establishing the existence of q and r such that:
$$a = dq + r \quad \text{with} \quad 0 \le r < d$$
Consider the set:
$$S = \{a - dn \mid n \in Z \text{ and } a - dn \ge 0\} \quad (*)$$
We first show that S is not empty:

If $a \ge 0$, then $a = a - d \cdot 0 \ge 0$, and therefore $a \in S$.
[0 is playing the role of n in (*)]

If $a < 0$, then $a - da \ge 0$, and therefore $a - da \in S$.
[a is playing the role of n in (*) and remember that $d \in Z^+$]

Since S is a nonempty subset of $\{0\} \cup Z^+$, it has a least element (Exercise 30, page 20); let's call it r. Since r is in S, there exists $q \in Z$ such that:
$$r = a - dq \quad (**)$$
To complete the existence part of the proof, we show that $r < d$.
Assume, to the contrary, that $r \ge d$. From:
$$r - d \underset{(**)}{=} (a - dq) - d = a - d(q+1)$$
we see that $r - d$ is of the form $a - dn$ (with $n = q + 1$). Moreover, our assumption that $r \ge d$ implies that $r - d \ge 0$. It follows that $r - d \in S$, **contradicting** the minimality of r.

To establish uniqueness, assume that:
$$a = dq + r \text{ with } 0 \le r < d \text{ and } a = dq' + r' \text{ with } 0 \le r' < d$$
[We will show that $q = q'$ and $r = r'$ (see margin)]

This is a common mathematical theme:
To establish that something is unique, consider two such "somethings" and then go on to show that the two "somethings" are, in fact, one and the same.

Since $r \ge 0$ and $r' < d$ (or $-r' > -d$): $r - r' \ge 0 - r' > 0 - d = -d$.
Since $r < d$ and $r' \ge 0$ (or $-r' \le 0$): $r - r' < d - r' \le d - 0 = d$
Thus: $-d < r - r' < d$, or $|r - r'| < d$
From $dq + r = dq' + r'$ we have: $r - r' = d(q' - q)$
(a multiple of d)

But if $|r - r'| < d$ and if $r - r'$ is a **multiple of d**, then $r - r' = 0$ (or $r = r'$). Returning to $dq + r = dq' + r'$ we now have:
$$dq + r = dq' + r' \Rightarrow dq = dq' \Rightarrow d(q - q') = 0 \underset{d \ne 0}{\Rightarrow} q = q'$$

EXAMPLE 1.8 Show that for any **odd** integer n, $8|(n^2-1)$.

SOLUTION: There are, at times, more than one way to stroke a cat:

<u>**Using Induction**</u>

We show that the proposition:
$$8|[(2m+1)^2 - 1]$$
holds for all $m \geq 0$ (**thereby covering all odd integers n**).

I. Valid at $m = 0$: $(2 \cdot 0 + 1)^2 - 1 = 0$.

II. Assume valid at $m = k$; that is:
$$(2k+1)^2 - 1 = 8t \text{ or } \mathbf{4k^2 + 4k = 8t}$$
for some integer t.

III. We are to establish validity at $m = k+1$; that is, that:
$$[2(k+1)+1]^2 - 1 = 8s$$
for some integer s. Let's do it:

$[2(k+1)+1]^2 - 1$
$= (2k+3)^2 - 1$
$= 4k^2 + 12k + 8$
$= \mathbf{(4k^2 + 4k)} + (8k+8)$
$= \mathbf{8t} + 8(k+1) = 8(t+k+1) = 8s$
↑
II

<u>**Using the Division Algorithm**</u>

We know that for any n there exists q such that:
$n = 2q$ or $n = 2q+1$ (*)
$n = 3q$ or $n = 3q+1$ or $n = 3q+2$ (**)
$\mathbf{n = 4q \text{ or } n = 4q+1 \text{ or } n = 4q+2 \text{ or } n = 4q+3}$

While (*) and (**) may not lead us to a fruitful conclusion, the bottom line does. Specifically:

For any n:
$n = 4q$ or $n = 4q+1$ or $n = 4q+2$ or $n = 4q+3$
If n is **odd**, then there are but the two possibilities:
$$n = 4q+1 \text{ or } n = 4q+3$$
We now show that, in either case $8|(n^2-1)$.

If $n = 4q+1$, then:
$$n^2 - 1 = (4q+1)^2 - 1 = 16q^2 + 8q + 1 - 1 = 8k$$
(with $k = 2q^2 + q$)

If $n = 4q+3$, then:
$$n^2 - 1 = (4q+3)^2 - 1 = 16q^2 + 24q + 9 - 1 = 8h$$
(with $h = 2q^2 + 3q + 1$)

CHECK YOUR UNDERSTANDING 1.10

Answer: See page A-5

Prove that for any integer n, $n^2 = 3q$ or $n^2 = 3q+1$ for some integer q.

DEFINITION 1.8
GREATEST COMMON DIVISOR

For given a and b not both zero, the **greatest common divisor** of a and b, denoted by $gcd(a,b)$, is the largest positive integer that divides both a and b.

THEOREM 1.6 If a and b are not both 0, then there exist s and t such that:
$$gcd(a, b) = sa + tb$$

PROOF: Let
$$G = \{x > 0 | x = ma + nb \text{ for some } m \text{ and } n\}$$
Assume, without loss of generality that $a \neq 0$. Since both a and $-a$ are of the form $ma + nb$: $a = 1a + 0b$ while $-a = (-1)a + 0b$; and since either a or $-a$ is positive: $G \neq \emptyset$. That being the case, the Well Ordering Principle (page 18) assures us that G has a **smallest element** $g = sa + tb$. We show that $g = gcd(a, b)$ by showing that (1): g divides both a and b, and that (2): every divisor of a and b also divides g.

(1) Applying the Division Algorithm we have:
$$a = \underset{(*)}{qg + r} \text{ with } \underset{(**)}{0 \leq r < g}.$$
Substituting $g = sa + tb$ in (*) brings us to:
$$a = q(sa + tb) + r$$
$$r = (1 - qs)a - tb$$
Since r is of the form $ma + nb$ with $r < g$, it cannot be in G, and must therefore be 0 [see (**)]. Consequently $a = qg$, and $g|a$. The same argument can be used to show that $g|b$.

(2) If $d|a$ and $d|b$, then, by Theorem 1.4(b) and (c), page 15: $d|g$.

CHECK YOUR UNDERSTANDING 1.11

Show that for any a and b not both zero:
$$gcd(a, b) = gcd(|a|, |b|).$$

Answer: See page A-5

DEFINITION 1.9   Two integers a and b, not both zero, are **relatively prime** if:
RELATIVELY PRIME
$$gcd(a, b) = 1$$

For example:
Since $gcd(15, 8) = 1$, 15 and 8 are relatively prime.
Since $gcd(15, 9) = 3 \neq 1$, 15 and 9 are not relatively prime.

THEOREM 1.7   Two integers, a and b, are relatively prime if and only if there exist $s, t \in \mathbb{Z}$ such that
$$1 = sa + tb$$

PROOF: To say that a and b are relatively prime is to say that $gcd(a, b) = 1$. The existence of integers s and t such that $1 = sa + tb$ follows from Theorem 1.6.

For the converse, assume that there exist integers s and t such that $1 = sa + tb$. Since $gcd(a, b)$ divides both a and b, it divides 1 [Theorem 1.4(b) and (c), page 15]; and, being positive, must equal 1.

THEOREM 1.8 Let $a, b, c \in Z$. If $a|bc$, and if $gcd(a, b) = 1$, then $a|c$.

PROOF: Let s and t be such that:
$$1 = sa + tb$$
Multiplying both sides of the above equation by c:
$$c = sac + tbc$$
Clearly $a|sac$. Moreover, since $a|bc$: $a|tbc$. The result now follows from Theorem 1.4(b), page 15.

CHECK YOUR UNDERSTANDING 1.12

Answer: See page A-5.

Let $a, b, c \in Z$. Show that if $a|bc$ and $a \nmid b$, then a and c can not be relatively prime.

PRIME NUMBERS

Chances are that you are already familiar with the important concept of a prime number; but just in case:

DEFINITION 1.10 An integer $p > 1$ is **prime** if 1 and p are its
PRIME only divisors.

For example: 2, 5, 7, and 11 are all prime, while 9 and 25 are not. Moreover, since any even number is divisible by 2, no even number greater than 2 is prime.

So, 2 is the oddest prime (sorry).

THEOREM 1.9 If p is prime and if $p|ab$, then $p|a$ or $p|b$.

PROOF: If $p|a$, we are done. We complete the proof by showing that if $p \nmid a$, then $p|b$:

Since the greatest common divisor of p and a divides p, it is either 1 or p. As it must also divide a, and since we are assuming $p \nmid a$, it must be that $gcd(p, a) = 1$. The result now follows from Theorem 1.8.

CHECK YOUR UNDERSTANDING 1.13

Answer: See page A-5.

Let p be prime. Use the Principle of Mathematical Induction to show that if $p|a_1 a_2 \cdots a_n$, then $p|a_i$ for some $1 \le i \le n$.

The following result is important enough to be called the Fundamental Theorem of Arithmetic.

THEOREM 1.10 Every integer n greater than 1 can be expressed uniquely (up to order) as a product of primes.

PROOF: We use *API* of page 16 (starting at $n = 2$) to establish the existence part of the theorem:

I. Being prime, 2 itself is already expressed as a product of primes.

II. Suppose a prime factorization exists for all m with $2 \leq m \leq k$.

III. We complete the proof by showing that $k + 1$ can be expressed as a product of primes:

If $k + 1$ is prime, then we are done.

If $k + 1$ is not prime, then $k + 1 = ab$, with $2 \leq a \leq b \leq k$. By our induction hypothesis, both a and b can be expressed as a product of primes. But then, so can $k + 1 = ab$.

For uniqueness, consider the set:

$$S = \{n \in Z^+ | n \text{ has two different prime decompositions}\}$$

Assume that $S \neq \emptyset$ (we will arrive at a contradiction).

The Well-Ordering Principle of page 17 assures us that S has a **smallest element**, let's call it **m**. Being in S, **m** has two distinct prime factorizations, say:

$$\boldsymbol{m} = p_1 p_2 \cdots p_s = q_1 q_2 \cdots q_t$$

Since $p_1 | p_1 p_2 \cdots p_s$ and since $p_1 p_2 \cdots p_s = q_1 q_2 \cdots q_t$ we have $p_1 | q_1 q_2 \cdots q_t$. By CYU 1.13, $p_1 | q_j$ for some $1 \leq j \leq t$.

Without loss of generality, let us assume that $p_1 | q_1$. Since q_1 is prime, its only divisors are 1 and itself. It follows, since $p_1 \neq 1$, that $p_1 = q_1$. Consequently:

$$p_1 p_2 \cdots p_s = q_1 q_2 \cdots q_t \Rightarrow p_1 p_2 \cdots p_s = p_1 q_2 \cdots q_t$$

$$\Rightarrow p_1 p_2 \cdots p_s - p_1 q_2 \cdots q_t = 0$$

$$\Rightarrow p_1(p_2 \cdots p_s - q_2 \cdots q_t) = 0$$

$$p_1 \neq 0: \Rightarrow p_2 \cdots p_s - q_2 \cdots q_t = 0$$

$$\Rightarrow p_2 \cdots p_s \underset{\uparrow}{=} q_2 \cdots q_t$$

two distinct prime decompositions for an integer smaller than **m** — contradicting the minimality on **m** in S

THEOREM 1.11 There are infinitely many primes.

PROOF: Assume that there are but a finite number of primes, say $S = \{p_1, p_2, \ldots, p_n\}$, and consider the number:

$$m = p_1 p_2 \cdots p_n + 1$$

Since $m \notin S$, it is not prime. By Theorem 1.10, some prime must divide m. Let us assume, without loss of generality, that $p_1 | m$. Since p_1 divides both m and $p_1 p_2 \ldots p_n$: $p_1 | [m - (p_1 p_2 \ldots p_n)]$ [Theorem 1.4(b), page 15]. A contradiction, since $m - (p_1 p_2 \ldots p_n) = 1$.

CHECK YOUR UNDERSTANDING 1.14

Answer: See page A-5.

Let a and b be relatively prime. Prove that if $a|n$ and $b|n$, then $ab|n$.

	EXERCISES	

Exercises 1-3. For given a and d, determine integers q and r, with $0 \leq r < d$, such that $a = dq + r$.

1. $a = 0, d = 1$
2. $a = -5, d = 133$
3. $a = -134, d = 5$

Exercises 4-6. Find the greatest common divisor of a and b.

4. $a = 120, b = 880$
5. $a = -10, b = 55$
6. $a = -134, b = 5$

Exercises 7-10. The **least common multiple** of nonzero integers $a_1, a_2, \ldots, a_n$, written $\text{lcm}(a_1, a_2, \ldots, a_n)$, is the smallest positive integer that is a multiple of each a_i; i.e. is divisible by each a_i. Find:

7. $\text{lcm}(12, 20)$
8. $\text{lcm}(3, 5, 9)$
9. $\text{lcm}(2, 3, 9, 15)$
10. $\text{lcm}(-3, 2, 4, 21)$

11. Let $a = p_1^{e_1} \cdot p_2^{e_2} \ldots p_n^{e_n}$ and $a = p_1^{f_1} \cdot p_2^{f_2} \ldots p_n^{f_n}$, where the p_is are distinct primes and where $e_i \geq 0$ and $f_i \geq 0$ for all i. Let $m_i = \min(e_i, f_1)$ (the smaller of the two numbers), and $M_i = \max(e_i, f_1)$ (the larger of the two numbers). Prove that:

    (a) $\gcd(a, b) = p_1^{m_1} \cdot p_2^{m_2} \ldots p_n^{m_n}$
    (b) $\text{lcd}(a, b) = p_1^{M_1} \cdot p_2^{M_2} \ldots p_n^{M_n}$ (see Exercise 7-10)

12. Prove that if 3 does not divide n, then $n = 3k + 1$ or $n = 3k + 2$ for some $k \in Z$.

13. Let n be such that $3 \nmid (n^2 - 1)$. Show that $3 | n$.

14. Show that if n is not divisible by 3, then $n^2 = 3m + 1$ for some integer m.

15. Show that an odd prime p divides $2n$ if and only if p divides n.

16. Prove that if $a = 6n + 5$ for some n, then $a = 3m + 2$ for some m.

17. Show that $2 | (n^4 - 3)$ if and only if $4 | (n^2 + 3)$.

18. Prove that any two consecutive odd positive integers are relatively prime.

19. Let a and b not both be zero. Prove that there exist integers s and t such that $n = sa + tb$ if and only if n is a multiple of $\gcd(a, b)$.

20. Prove that the only three consecutive odd numbers that are prime are 3, 5, and 7.

21. Show that a prime p divides n^2 if and only if p divides n.

22. Prove that every odd prime p is of the form $4n + 1$ or of the form $4n + 3$ for some n.

23. Prove that every prime $p > 3$ is of the form $6n + 1$ or of the form $6n + 5$ for some n.

24. Prove that every prime $p > 5$ is of the form $10n + 1$, $10n + 3$, $10n + 7$, or $10n + 9$ for some n.

25. Prove that a prime p divides $n^2 - 1$ if and only if $p|(n-1)$ or $p|(n+1)$.

26. Prove that every prime of the form $3n + 1$ is also of the form $6k + 1$.

27. Prove that if n is a positive integer of the form $3k + 2$, then n has a prime factor of this form as well.

28. Prove that $a > 1$ and $b > 1$ are relatively prime if and only if no prime in the prime decomposition of a appears in the prime decomposition of b.

29. Prove that if the integer $n > 1$ satisfies the property that if $n|ab$, then $n|a$ or $n|b$ for every pair of integers a and b, then n is prime.

30. Prove that $n > 1$ is prime if and only if n is not divisible by any prime p with $p \leq \sqrt{n}$.

	PROVE OR GIVE A COUNTEREXAMPLE	

31. There exists an integer n such that $n^2 = 3m - 1$ for some m.
32. If $a = 3m + 2$ for some m, then $a = 6n + 5$ for some n.
33. If m and n are odd integers, then either $m + n$ or $m - n$ is divisible by 4.
34. For any a, and b not both 0, there exist a unique pair of integers s and t such that $gcd(a, b) = s \cdot a + t \cdot b$.
35. For every n, $3|(4^n - 1)$.
36. For every $n \in Z^+$, $3|(4^n + 1)$.

§4. EQUIVALENCE RELATIONS

Recall that $X \times Y$, called the **Cartesian Product** of X with Y, is the set of all ordered pairs (x, y), with $x \in X$ and $y \in Y$.

In Section 2 we defined a function from a set X to a set Y to be a subset $f \subseteq X \times Y$ such that:

For every $x \in X$ there exists a unique $y \in Y$ with $(x, y) \in f$.

Removing all restrictions, we arrive at a far more general concept than that of a function:

DEFINITION 1.11
RELATION

A **relation** E **from a set** X **to a set** Y is any subset $E \subseteq X \times Y$.
A relation from a set X to X is said to be a **relation on** X.

Each and every subset of $\mathfrak{R} \times \mathfrak{R}$, including the chaotic one in the margin, is a relation on $\mathfrak{R}$, suggesting that Definition 1.11 is a tad too general. Some restrictions are in order:

DEFINITION 1.12

A relation E on a set X is a subset $E \subseteq X \times X$ and is said to be:

REFLEXIVE

Reflexive: $(x, x) \in E$ for every $x \in X$.
(Every element of X is related to itself)

SYMMETRIC

Symmetric: If $(x, y) \in E$ then $(y, x) \in E$.
(If x is related to y, then y is related to x)

TRANSITIVE

Transitive: If $(x, y) \in E$ and $(y, z) \in E$ then $(x, z) \in E$.
(If x is related to y, and y is related to z, then x is related to z)

EQUIVALENCE RELATION

An **equivalence relation** on a set X is a relation that is reflexive, symmetric and transitive.

The notation $x{\sim}y$ is often used to indicate that x is related to y with respect to some understood relation E. Utilizing that option, we can rephrase Definition 1.12 as follows:

An **equivalence relation** $\sim$ on a set X is a relation which is

Reflexive: if $x{\sim}x$ for every $x \in X$.

Symmetric: if $x{\sim}y$, then $y{\sim}x$.

and Transitive: if $x{\sim}y$ and $y{\sim}z$, then $x{\sim}z$.

EXAMPLE 1.9

Show that the relation $\frac{a}{b}{\sim}\frac{c}{d}$ if $ad = bc$ is an equivalence relation on the set of rational numbers.

As you know, when it comes to rational numbers, one simply writes $\frac{2}{3} = \frac{4}{6}$ rather than $\frac{2}{3} \sim \frac{4}{6}$.

SOLUTION:

Reflexive: $\frac{a}{b} \sim \frac{a}{b}$ since $ab = ba$.

Symmetric: $\frac{a}{b} \sim \frac{c}{d} \Rightarrow ad = bc \Rightarrow cb = da \Rightarrow \frac{c}{d} \sim \frac{a}{b}$.

Transitive: $\frac{a}{b} \sim \frac{c}{d}$ and $\frac{c}{d} \sim \frac{e}{f} \Rightarrow \underset{(*)}{ad = bc}$ and $\underset{(**)}{cf = de}$

We establish the fact that $\frac{a}{b} \sim \frac{e}{f}$ by showing that $\mathbf{af = be}$:

$$af = \underset{\text{see }(*)}{\frac{bc}{d}} \cdot f = \underset{\text{see }(**)}{\frac{bc}{d}} \cdot \frac{de}{c} = be$$

Recall that $a|b$ means that a divides b (see Definition 1.7, page 15).

EXAMPLE 1.10 Show that the relation $a \sim b$ if $2|(3a-b)$ is an equivalence relation on Z.

SOLUTION: The relation $a \sim b$ if $2|(3a-b)$ is:

Reflexive. $a \sim a$, since: $3a - a = 2a$
                                         ↑ — here, a is playing the role of b

Symmetric. Assume that $a \sim b$, which is to say, that:
$$3a - b = 2h \text{ for some } h \in Z \quad (*)$$
We are to show that $b \sim a$, which is to say, that:
$$3b - a = 2n \text{ for some } n \in Z$$
Lets do it. From (*) $b = 3a - 2h$.

Hence: $3b - a = 3(3a - 2h) - a = 2(\mathbf{4a - 3h}) = \mathbf{2n}$

An expression of the form $a = \frac{2h+b}{3}$ is unacceptable in the solution process, since we are involved with the set Z of integers and not "fractions."

TRANSITIVE: Assume that $a \sim b$ and $b \sim c$; which is to say, that:
(1) $3a - b = 2h$ and (2) $3b - c = 2k$ for $h, k \in Z$

We are to show that $a \sim c$; which is to say, that: $3a - c = 2n$.

Let's do it. From (2): $c = 3b - 2k$.

Hence: $3a - c = 3a - (3b - 2k)$

From (1): $= 2h + b - (3b - 2k) = 2(\mathbf{h + k - b}) = \mathbf{2n}$

CHECK YOUR UNDERSTANDING 1.15

Two sets A and B are said to have the same **cardinality**, written $\text{Card}(A) = \text{Card}(B)$, if there exists a bijection $f: A \to B$.

Show that the relation $A \sim B$ if $\text{Card}(A) = \text{Card}(B)$ is an equivalence relation on any collection S of sets.

Note: In a sense, the term "*same cardinality*" can be interpreted to mean "*same number of elements.*" The classier terminology is used since the expression "same number of elements" suggests that we have associated a number to each set, even those that are infinite. A further discussion on cardinality if offered in the exercises.

Answer: See page A-6

DEFINITION 1.13

EQUIVALENCE CLASS

Let ~ be an equivalence relation on X. For each $x_0 \in X$, the **equivalence class** of x_0, denoted by $[x_0]$, is the set:

$$[x_0] = \{x \in X | x \sim x_0\}$$

In words: The equivalence class of x_0 consists of all elements of X that are related to x_0. We now show that any element in $[x_0]$ will generate the same equivalence class:

THEOREM 1.12

Let ~ be an equivalence relation on X. For any $x_1, x_2 \in X$:

$$x_1 \sim x_2 \Leftrightarrow [x_1] = [x_2]$$

PROOF: Assume that $x_1 \sim x_2$. We show that $[x_1] \subseteq [x_2]$ (a similar argument can be used to show that $[x_2] \subseteq [x_1]$ and that therefore $[x_1] = [x_2]$):

$$x \in [x_1] \Rightarrow x \sim x_1$$

By transitivity, since $x_1 \sim x_2$: $x \sim x_2 \Rightarrow x \in [x_2]$

Conversely, if $[x_1] = [x_2]$, then, since $x_1 \in [x_2]$: $x_1 \sim x_2$.

EXAMPLE 1.11

Determine the set $\{[n]\}_{n \in Z}$ of equivalence classes corresponding to the equivalence relation $a \sim b$ if $2|(3a-b)$ of Example 1.10.

SOLUTION: Let's start off with $a = 0$. By definition:

$[0] = \{b \in Z | 2|(-b)\} = \{2n | n \in Z\}$ (the even integers)

Since 1 is not in $[0]$, $[1]$ will differ from $[0]$ (Theorem 1.12). Specifically:

$[1] = \{b \in Z | 2|(1-b)\} = \{2n+1 | n \in Z\}$ (the odd integers)

In the above example the give equivalence relation decomposed Z into disjoint equivalence classes; namely:

$$Z = [0] \cup [1] = \{\text{even integers}\} \cup \{\text{odd integers}\}$$

To put it another way: the equivalence classes in Example 1.11 effected a partition of Z, where:

To put it roughly: A partition of a set S chops S up into disjoint pieces.

DEFINITION 1.14
PARTITION

A set of nonempty subsets $\{S_\alpha\}_{\alpha \in A}$ of a set X is said to be a **partition** of X if:

(i) $X = \bigcup_{\alpha \in A} S_\alpha$

(ii) If $S_\alpha \cap S_{\bar\alpha} \neq \varnothing$ then $S_\alpha = S_{\bar\alpha}$

In the above, $\{S_\alpha\}_{\alpha \in A}$ is being indexed by the set A, as is the case with the union $\bigcup_{\alpha \in A} S_\alpha$. In particular: If $A = \{1, 2\}$, then:

$$\{S_\alpha\}_{\alpha \in A} = \{S_\alpha\}_{\alpha \in \{1,2\}} = \{S_1, S_2\} \text{ and } \bigcup_{\alpha \in A} S_\alpha = \bigcup_{\alpha \in \{1,2\}} S_\alpha = S_1 \cup S_2$$

And, if $A = Z^+ = \{1, 2, 3, \ldots\}$, then:

$$\{S_\alpha\}_{\alpha \in A} = \{S_i\}_{i \in Z^+} = \{S_i\}_{i=1}^\infty \text{ and } \bigcup_{\alpha \in Z^+} S_\alpha = \bigcup_{i=1}^\infty S_i$$

Figure 1.5(a) displays a 5-subset partition $\{S_1, S_2, S_3, S_4, S_5\}$ of the indicated set. An infinite partition of $[0, \infty)$ is represented in Figure 1.5(b): $\{[n, n+1)\}_{n=0}^\infty$

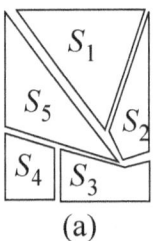
(a)

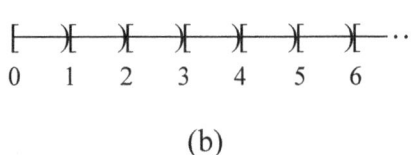
(b)

Figure 1.5

CHECK YOUR UNDERSTANDING 1.16

Determine if the given collection of subsets of $\Re$ is a partition of $\Re$?

(a) $\{[n, n+1]\}_{n \in Z}$

(b) $\{\{n\} | n \in Z\} \cup \{(i, i+1)\}_{i=0}^\infty \cup \{(-i-1, -i)\}_{i=0}^\infty$

(a): No (b): Yes

There is an important connection between the equivalence relations on a set X and the partitions of X, and here it is:

THEOREM 1.13

(a) If $\sim$ is an equivalence relation on X, then the set of its equivalence classes, $\{[x]\}_{x \in X}$, is a partition of X.

(b) If $\{S_\alpha\}_{\alpha \in A}$ is a partition of X, then the relation $x_1 \sim x_2$ if $x_1, x_2 \in S_\alpha$ for some $\alpha \in A$ is an equivalence relation on X.

PROOF: (a) We Show that:

(i) $X = \bigcup_{x \in X} [x]$

and (ii) If $[x_1] \cap [x_2] \neq \emptyset$, then $[x_1] = [x_2]$.

(i): Since ~ is an equivalence relation, $x \sim x$ for every $x \in X$. It follows that $x \in [x]$ for every $x \in X$, and that therefore $X = \bigcup_{x \in X} [x]$.

(ii): If $[x_1] \cap [x_2] \neq \emptyset$, then there exists $x_0 \in [x_1] \cap [x_2]$.

Since $x_0 \in [x_1]$ and $x_0 \in [x_2]$: $x_0 \sim x_1$ and $x_0 \sim x_2$.

By symmetry and transitivity: $x_1 \sim x_2$

By Theorem 1.12: $[x_1] = [x_2]$

(b) Let $\{S_\alpha\}_{\alpha \in A}$ be a partition of X. We show that the relation:

$x_1 \sim x_2$ if there exists $\alpha \in A$ such that $x_1, x_2 \in S_\alpha$ is an equivalence relation on X:

Reflexive: To say that $x \sim x$, is to say that x belongs to the same S_α as itself, and it certainly does.

Symmetric: $x \sim y \Rightarrow \exists \alpha \in A \ni x, y \in S_\alpha \Rightarrow y, x \in S_\alpha \Rightarrow y \sim x$

Transitive: Assume $x \sim y$ and $y \sim z$. We show that $x \sim z$:

Since $x \sim y$: $x, y \in S_\alpha$ for some $\alpha \in A$.

Since $z \sim y$: $y, z \in S_{\bar\alpha}$ for some $\bar\alpha \in A$.

Since $S_\alpha \cap S_{\bar\alpha} \neq \emptyset$ (y is contained in both sets): $S_\alpha = S_{\bar\alpha}$.

It follows that both x and z are in S_α (or in $S_{\bar\alpha}$ if you prefer), and that, consequently: $x \sim z$.

CONGRUENCE MODULO n

Here is a particularly important equivalence relation of the set of integers:

THEOREM 1.14 Let $n \in Z^+$. The relation $a \sim b$ if $n | (a - b)$ is an equivalence relation on Z.

PROOF: Reflexive: $a \sim a$ since $n | (a - a)$.

Symmetric: $a \sim b \Rightarrow n | (a - b) \Rightarrow n | (b - a) \Rightarrow b \sim a$.

Transitive:

$$a \sim b \text{ and } b \sim c \Rightarrow n|(a-b) \text{ and } n|(b-c)$$

Theorem 1.4(b), page 15: $\Rightarrow n|[(a-b)+(b-c)]$

$$\Rightarrow n|(a-c) \Rightarrow a \sim c$$

> In the event that $n|(a-b)$, we say that:
> **a is congruent to b modulo n** and write $a \equiv b \bmod n$

THEOREM 1.15 Let $n \in Z^+$. If $a \equiv \bar{a} \bmod n$ and $b \equiv \bar{b} \bmod n$, then:

(a) $a + b \equiv \bar{a} + \bar{b} \bmod n$

(b) $ab \equiv \bar{a}\bar{b} \bmod n$

PROOF: (a) If $n|(a-\bar{a})$ and $n|(b-\bar{b})$, then:

$$n|[(a-\bar{a})+(b-\bar{b})] \Rightarrow n|[(a+b)-(\bar{a}+\bar{b})]$$

(a) If $n|(a-\bar{a})$ and $n|(b-\bar{b})$, then:

(1) $a - \bar{a} = hn$ and (2) $b - \bar{b} = kn$ for $h, k \in Z$

We are to show that $n|(ab - \bar{a}\bar{b})$; which is to say that $ab - \bar{a}\bar{b} = ns$

Lets do it: $ab - \bar{a}\bar{b} = (ab - \bar{a}b) + (\bar{a}b - \bar{a}\bar{b})$

$$= (a - \bar{a})b + \bar{a}(b - \bar{b})$$

$$= hnb + \bar{a}kn = n(\mathbf{hb + \bar{a}k}) = n\mathbf{s}$$

CHECK YOUR UNDERSTANDING 1.17

Let $n \in Z^+$. Let $a = d_a n + r_a$ and $b = d_b n + r_b$ with $0 \le r_a < n$ and $0 \le r_b < n$ (see Theorem 1.5, page 21). Prove that:

$$a \equiv b \bmod n \text{ if and only if } r_a = r_b$$

(same remainder when dividing by n)

Answer: See page A-6

Theorem 1.13 assures is that the equivalent classes associated with the equivalence relation of Theorem 1.15 partition the set of integers. Focusing on $n = 5$, we see that the equivalence class containing 0 consists of all multiples of 5, as the remainder of any multiple of 5 when divided by 5, is the same as that obtained by dividing 0 by 5 (see CYU 1.17). Specifically:

$$[0]_5 = \{\ldots, -20, -15, -10, -5, 0, 5, 10, 15, 20, \ldots\}$$

Note that the above equivalence class has many "names". It can also, be called the equivalent class containing 235, among infinitely many other choices:

$$[0]_5 = [125]_5 = [-15]_5 = \cdots$$

The same can be said about the four remaining equivalence classes:

$$[1]_5 = \{\ldots, -14, -9, -4, 1, 6, 11, 16, \ldots\}$$
$$[2]_5 = \{\ldots, -13, -8, -3, 2, 7, 12, 17, \ldots\}$$
$$[3]_5 = \{\ldots, -12, -7, -2, 3, 8, 12, 18, \ldots\}$$
$$[4]_5 = \{\ldots, -11, -6, -1, 4, 9, 13, 19, \ldots\}$$

Note that $[5] = [0]$.

Can we define a sum on the above five equivalence classes? Yes:

$$[a]_5 [+] [b]_5 = [a+b]_5$$

The above sum is well defined, in that it is independent of the chosen representatives in the two equivalence classes. Indeed:

THEOREM 1.16 For given $n \in Z^+$, let $[Z]_n$ denote the set of equivalence classes associated with the equivalence relation $a \sim b$ if $n|(a-b)$; i.e:

$$[Z]_n = \{[0]_n, [1]_n, \ldots, [n-1]_n\}$$

Then:

(a) For any $[a]_n, [b]_n \in [Z]_n$, the operation

$$[a]_n [+] [b]_n = [a+b]_n$$

is well defined.

(b) For any $[a]_n, [b]_n, [c]_n \in [Z]_n$:

$$([a]_n [+] [b]_n) [+] [c]_n = [a]_n [+] ([b]_n [+] [c]_n)$$

(associative property)

PROOF: (a) We show that if $[a]_n = [\bar{a}]_n$ and $[b]_n = [\bar{b}]_n$, then $[a+b]_n = [\bar{a}+\bar{b}]_n$ (i.e the sum is independent of the chosen representatives for the equivalence classes $[a]_n$ and $[b]_n$):

$$[a]_n = [\bar{a}]_n \Rightarrow n|(a-\bar{a}) \Rightarrow a-\bar{a} = hn, \text{ for } h \in Z$$

and: $[b]_n = [\bar{b}]_n \Rightarrow n|(b-\bar{b}) \Rightarrow b-\bar{b} = kn$, for $k \in Z$.

Since $(a+b)-(\bar{a}+\bar{b}) = (a-\bar{a})-(b-\bar{b}) = (h-k)n$:

$$[a+b]_n = [\bar{a}+\bar{b}]_n$$

(b) $([a]_n [+] [b]_n) [+] [c]_n = [a+b]_n [+] [c]_n = [(a+b)+c]_n$
$$= [a+(b+c)]_n$$
$$= [a]_n [+] ([b]_n [+] [c]_n)$$

CHECK YOUR UNDERSTANDING 1.18

(a) Verify that the product $[a]_n[b]_n = [ab]_n$ in Z_n is well defined. That is: if $[a]_n = [\bar{a}]_n$ and $[b]_n = [\bar{b}]_n$, then: $[ab]_n = [\bar{a}\bar{b}]_n$.

(b) Prove that $[a]_n([b]_n[c]_n) = ([a]_n[b]_n)[c]_n$

(c) Prove that $[a]_n([b]_n[+][c]_n) = [a]_n[b_n][+][a]_n[c]_n$.

1.4 Equivalence Relations

	EXERCISES	

Exercises 1-3. Show that the given relation is an equivalence relation on Z.
1. $a \sim b$ if $|a| = |b|$.   2. $a \sim b$ if $2 | (a - 3b)$.   3. $a \sim b$ if $5 | (a - b)$.

Exercises 4-7. Show that the given relation is an equivalence relation on Q, the set of rational numbers.

4. $\dfrac{a}{b} \sim \dfrac{c}{d}$ if $\dfrac{a}{b} - \dfrac{c}{d} \in Z$.

5. $\dfrac{a}{b} \sim \dfrac{c}{d}$ if $2 | (b + d)$.

6. $\dfrac{a}{b} \sim \dfrac{c}{d}$ if $(ad - bc)(b^2 + d^2) = 0$.

7. $\dfrac{a}{b} \sim \dfrac{c}{d}$ if $(ad)^2 - (bc)^2 = 0$.

Exercises 8-13. Show that the given relation is an equivalence relation on $\Re$.
8. $x \sim y$ if $x^2 = y^2$.   9. $x \sim y$ if $|x| = |y|$.   10. $x \sim y$ if $|x + 1| = |y + 1|$.
11. $x \sim y$ if $x - y \in Z$.   12. $x \sim y$ if $\sin x = \sin(y + 2\pi)$.   13. $x \sim y$ if $x^2 - y^2 = 0$.

Exercises 14-17. Show that the given relation is an equivalence relation on $\Re^2$.
14. $(x_0, y_0) \sim (x_1, y_1)$ if $x_0 + y_0 = x_1 + y_1$.   15. $(x_0, y_0) \sim (x_1, y_1)$ if $x_0 y_0 = x_1 y_1$.
16. $(x_0, y_0) \sim (x_1, y_1)$ if $x_0^2 + y_0^2 = x_1^2 + y_1^2$.   17. $(x_0, y_0) \sim (x_1, y_1)$ if $x_0 = x_1$.

Exercises 18-21. Show that the given relation is not an equivalence relation on $\Re^2$.
18. $(x_0, y_0) \sim (x_1, y_1)$ if $x_0 = y_1$.   19. $(x_0, y_0) \sim (x_1, y_1)$ if $x_0 - y_1 = y_0 - x_1$.
20. $(x_0, y_0) \sim (x_1, y_1)$ if $x_0 x_1 = y_0 y_1$.   21. $(x_0, y_0) \sim (x_1, y_1)$ if $x_0 x_1 \leq 0$ and $y_0 y_1 \leq 0$.

Exercises 22-30. Determine whether or not the given relation is an equivalence relation on $\Re^3$.
22. $(x_0, y_0, z_0) \sim (x_1, y_1, z_1)$ if $y_0 = y_1$.
23. $(x_0, y_0, z_0) \sim (x_1, y_1, z_1)$ if $x_0 + y_0 + z_0 = x_1 + y_1 + z_1$.
24. $(x_0, y_0, z_0) \sim (x_1, y_1, z_1)$ if $x_0 = y_1 + z_1$.
25. $(x_0, y_0, z_0) \sim (x_1, y_1, z_1)$ if $x_0 z_0 + 2 y_0 \leq x_1 z_1 + 2 y_1$.
26. $(x_0, y_0, z_0) \sim (x_1, y_1, z_1)$ if $x_0 + 2 y_0 - 3 z_0 = x_1 + 2 y_1 - 3 z_1$.
27. $(x_0, y_0, z_0) \sim (x_1, y_1, z_1)$ if $(x_0 + y_0 + z_0)^2 = (x_1 + y_1 + z_1)^2$.
28. $(x_0, y_0, z_0) \sim (x_1, y_1, z_1)$ if $x_0^2 + y_0^2 + z_0^2 = x_1^2 + y_1^2 + z_1^2$.
29. $(x_0, y_0, z_0) \sim (x_1, y_1, z_1)$ if $x_0 + y_0 + z_0 + x_1 + y_1 + z_1 \geq 0$.
30. $(x_0, y_0, z_0) \sim (x_1, y_1, z_1)$ if $|y_0 z_0| = |y_1 z_1|$.

Exercises 31-34. Determine whether or not the given relation is an equivalence relation on $M_{2\times 2}$.

31. $\begin{bmatrix} a & b \\ c & d \end{bmatrix} \sim \begin{bmatrix} \bar{a} & \bar{b} \\ \bar{c} & \bar{d} \end{bmatrix}$ if $a = \bar{d}$.

32. $\begin{bmatrix} a & b \\ c & d \end{bmatrix} \sim \begin{bmatrix} \bar{a} & \bar{b} \\ \bar{c} & \bar{d} \end{bmatrix}$ if $abc = \bar{a}\bar{b}\bar{c}$.

33. $\begin{bmatrix} a & b \\ c & d \end{bmatrix} \sim \begin{bmatrix} \bar{a} & \bar{b} \\ \bar{c} & \bar{d} \end{bmatrix}$ if $ad - bc = \bar{a}\bar{d} - \bar{b}\bar{c}$.

34. $\begin{bmatrix} a & b \\ c & d \end{bmatrix} \sim \begin{bmatrix} \bar{a} & \bar{b} \\ \bar{c} & \bar{d} \end{bmatrix}$ if $ad - \bar{b}\bar{c} = \bar{a}\bar{d} - bc$.

Exercises 35-41. Show that the given relation is an equivalence relation on $F(Z) = \{f: Z \to Z\}$ (the set of functions from Z to Z).

35. $f \sim g$ if $f(1) = g(1)$.
36. $f \sim g$ if $|f(n)| = |g(n)|$ for every $n \in Z$.
37. $f \sim g$ if $|f(n)| = |g(n)|$ for every $n \in Z$.
38. $f \sim g$ if $2 | [f(n) + g(n)]$ for every $n \in Z$.
39. $f \sim g$ if $f(n + m) = g(n + m)$ for every $n, m \in Z$.
40. $f \sim g$ if $3 | (2f(n) + g(n))$ for every $n \in Z$.
41. $f \sim g$ if $3 | [2(g \circ f)(n) + f(n)]$ for every $n \in Z$.

Exercises 42-47. Describe the set of equivalence classes for the equivalence relation of:

42. Exercise 1
43. Exercise 3
44. Exercise 5
45. Exercise 9
46. Exercise 15
47. Exercise 17

Exercises 48-52. Show that the given collection S of subsets of the set X is a partition of X.

48. $X = \Re$, $S = \{(-\infty, 0) \cup \{0\} \cup (0, \infty)\}$.
49. $X = Z$, $S = \{\{3n | n \in Z\} \cup \{3n + 1 | n \in Z\} \cup \{3n + 2 | n \in Z\}\}$.
50. $X = Z^+ \times Z^+$, $S = \{(a, b) | \gcd(a, b) = n\}_{n \in Z^+}$.
51. $X = \Re \times \Re$, $S = \{(x, y) | y = x + b\}_{b \in R}$.
52. $X = \Re \times \Re$, $S = \{(x, y) | x^2 + y^2 = r^2\}_{r \in \Re}$.

Exercises 53-54. (Congruences) Let $n \in Z^+$. Use the Principle of Mathematical Induction to show that:

53. If $a_i \equiv \bar{a}_i \mod n$ for $1 \leq i \leq m$, then $a_1 + a_2 + \cdots + a_m \equiv \bar{a}_1 + \bar{a}_2 + \cdots + \bar{a}_m \mod n$.

54. If $a_i \equiv \bar{a}_i \mod n$ for $1 \leq i \leq m$, then $a_1 a_2 \cdots a_m \equiv \bar{a}_1 \bar{a}_2 \cdots \bar{a}_m \mod n$.

55. Show that the relation $A \sim B$ if $\text{Card}(A) = \text{Card}(B)$ is an equivalence relation on $P(X)$ for any set X. Suggestion: Consider Theorem 1.1, page 5.

	PROVE OR GIVE A COUNTEREXAMPLE	

56. The union of any two equivalence relations on any given nonempty set X is again an equivalence relation on X.

57. The intersection of any two equivalence relations on any given nonempty set X is again an equivalence relation on X.

58. For $a, b, n, m \in Z^+$, let S_n and S_m denote the set of equivalence classes associated with the equivalence relations $a \sim b$ if $n|(a-b)$ and $a \sim b$ if $m|(a-b)$, respectively. If $n < m$, then $S_n \subset S_m$.

59. If $n \geq 2$, then every integer is congruent modulo n to exactly one of the integers $0 \leq m < n$.

60. If $C \subseteq X$, $A \sim B$ if $A \cap C = B \cap C$ is an equivalence relation on $P(X)$.

61. There exists an equivalence relation on the set $\{1, 2, 3, 4, 5\}$ for which each equivalence class contains an even number of elements.

Part 2
Groups

§1. DEFINITIONS AND EXAMPLES

The following properties reside in the familiar set Z of integers:

	Property	Example:
Closure	$a + b \in Z \quad \forall a, b \in Z$	$5 + 7 \in Z$
Associative	1. $a + (b + c) = (a + b) + c \quad \forall a, b \in Z$	$5 + (4 + 1) = (5 + 4) + 1$
Identity	2. $a + 0 = a \quad \forall a \in Z$	$4 + 0 = 4$
Inverse	3. $a + (-a) = 0 \quad \forall a \in Z$	$5 + (-5) = 0$

A generalization of the above properties bring us to the definition of a group — an abstract structure upon which rests a rich theory, with numerous applications throughout mathematics, the sciences, architecture, music, the visual arts, and elsewhere:

> A **binary operator** on a set X is a function that assigns to any **two** elements in X an element **in** X. Since the function value resides back in X, one says that the operator is **closed**.

> Evariste Galois defined the concept of a group in 1831 at the age of 20. He was killed in a duel one year later, while attempting to defend the honor of a prostitute.

DEFINITION 2.1
GROUP

A **group** $\langle G, * \rangle$, or simply G, is a nonempty set G together with a binary operator, $*$, (see margin) such that:

Associative Axiom: 1. $a*(b*c) = (a*b)*c$ for every $a, b, c \in G$.

Identity Axiom: 2. There exists an element in G, which we will label e, such that $a*e = a$ for every $a \in G$.

Inverse Axiom: 3. For every $a \in G$ there exists an element, $a' \in G$ such that $a*a' = e$.

In particular, $\langle Z, + \rangle$ is a group; with "+, 0, and $-a$" playing the role of "$*$, e, and a'" in the above definition.

> We show, in the next section, that both the **identity element** e and the **inverse element** a' of Axioms 2 and 3 are, in fact, both unique and "ambidextrous:"
> $a*e = e*a = a$
> $a*a' = a'*a = e$

Is the set of integers under multiplication a group? No:

While "regular" multiplications is an associative binary operator on Z, with 1 as identity, no integer other than ± 1 has a multiplicative inverse **in Z**.

> Yes, there is a number whose product with 2 is 1:
> $2 \cdot \frac{1}{2} = 1$, but $\frac{1}{2} \notin Z$.

Bottom line: The set of integers under multiplication is **not** a group.

CHECK YOUR UNDERSTANDING 2.1

Determine if the given set is a group under the given operation. If not, specify which of the axioms of Definition 2.1 do not hold.

(a) The set Q of rational numbers under addition.

(b) The set $\Re$ of real numbers under addition.

(c) The set $\Re$ of real numbers under multiplication.

(d) The set $\Re^+ = \{r \in \Re | r > 0\}$ of positive real numbers under multiplication.

(a), (b), and (d) are groups. (c) is not a group.

We now move Theorem 1.16 of page 35 up a notch:

THEOREM 2.1 For given $n \in Z^+$, let $[Z]_n$ denote the set of equivalence classes associated with the equivalence relation $a \sim b$ if $n | (a-b)$; i.e:

$$[Z]_n = \{[0]_n, [1]_n, ..., [n-1]_n\}$$

Then: $\langle [Z]_n, [+] \rangle$ with $[a]_n [+] [b]_n = [a+b]_n$

is a group

PROOF: We already know that $[+]$ is a well defined associative operator. The identity and inverse axioms of Definition 2.1 are also met:

Identity: For any $[a]_n \in [Z]_n$: $[a]_n [+] [0]_n = [a+0]_n = [a]_n$.

Inverses: For any $[a]_n \in [Z]_n$: $[a]_n [+] [-a]_n = [a-a]_n = [0]_n$

Molding Theorem 2.1 into a more compact form by replacing each equivalent class $[a]_n$ with the smallest nonnegative integer in that class, we come to:

THEOREM 2.2 For given $n \in Z^+$, let $Z_n = \{0, 1, 2, ..., n-1\}$, and let $a +_n b = r$, where $a + b = dn + r$.

Then $\langle Z_n, +_n \rangle$ is a group.

You are invited to formally establish this result in Exercise 51.

The above sum is called **addition modulo n.** Note that $a +_n 0 = a$ for every $a \in Z_n$, and that for any $a \in Z_n$: $a + (n-a) = 0$

For example, if $n = 5$ then $Z_5 = \{0, 1, 2, 3, 4\}$, and:

$$1 +_5 2 = 3, \quad 4 +_5 4 = 3, \quad \text{and} \quad 3 +_5 2 = 0$$
$$\uparrow$$
$$3 \equiv (4+4) \bmod 5$$

CHECK YOUR UNDERSTANDING 2.2

Complete the following (self-explanatory) **group table** for $\langle Z_4, +_4 \rangle$.

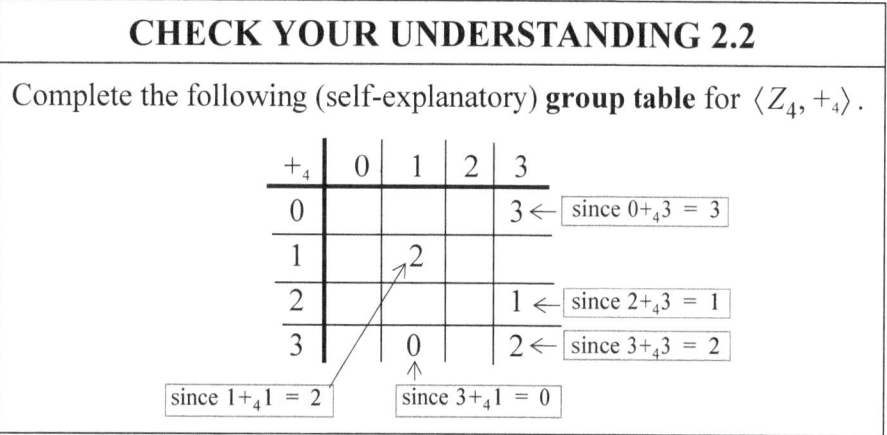

Answer: See page A-7.

Groups containing infinitely many elements, like $\langle Z, + \rangle$ and $\langle \Re, + \rangle$, are said to be **infinite groups**. Those containing finite may elements, like $\langle Z_n, +_n \rangle$ which contains n elements, are said to be **finite groups.**

DEFINITION 2.2
ORDER OF A GROUP

Let G be a finite group. The number of elements in G is called the **order of G**, and is denoted by $|G|$.

GROUP TABLES AND BEYOND

The group Z_4, with table depicted in Figure 2.1(a), has order 4. Another group of order 4, the so-called **Klein 4-group**, appears in Figure 2.1(b).

Z_4:

$+_4$	0	1	2	3
0	0	1	2	3
1	1	2	3	0
2	2	3	0	1
3	3	0	1	2

(a)

K:

*	e	a	b	c
e	e	a	b	c
a	a	e	c	b
b	b	c	e	a
c	c	b	a	e

(b)

Figure 2.1

Is K really a group? Well, the above table leaves no doubt that the closure and identity axioms are satisfied (e is the identity element). Moreover, each element has an inverse, namely itself: $ee = e, aa = e, bb = e$, and $cc = e$. Finally, though a bit tedious, you can check directly that the associative property holds [for example: $(ab)a = ca = b$ and $a(ba) = ac = b$]. You can also see that K is an abelian group; where:

Abelian groups are also said to be commutative groups.

DEFINITION 2.3
ABELIAN GROUP

A group $\langle G, * \rangle$ is **abelian** if
$$a*b = b*a \text{ for every } a, b \in G$$

We will soon show that Z_4 and K are the only groups of order 4, but first:

THEOREM 2.3 Every element of a finite group G must appear once and only once in each row and each column of its group table.

PROOF: Let $G = \{e, a_1, a_2, \ldots, a_{n-1}\}$. By construction, the ith row of G's group table is precisely $a_i e, a_i a_1, a_i a_2, \ldots, a_i a_{n-1}$. The fact that every element of G appears exactly one time in that row is a consequence of Exercise 50, which asserts that the function $f_{a_i}: G \to G$ given by $f_{a_i}(g) = a_i g$ is a bijection. As for the columns:

CHECK YOUR UNDERSTANDING 2.3

Answer: See page A-8.

Complete the proof of Theorem 2.3.

An alternative proof is offered in Appendix B, page B-1.

We now show that the two groups in Figure 2.1 represent all groups of order four. To begin with, we note that any group table featuring the four elements $\{e, a, b, c\}$ must "start off" as in T in Figure 2.2, for e represents the identity element.

T:	*	e	a	b	c
	e	e	a	b	c
	a	a	▧		
	b	b			
	c	c			

E:	*	e	a	b	c		B:	*	e	a	b	c		C:	*	e	a	b	c
	e	e	a	b	c			e	e	a	b	c			e	e	a	b	c
	a	a	e	▨				a	a	b	▨				a	a	c	▨	
	b	b						b	b						b	b			
	c	c						c	c						c	c			

Figure 2.2

Since no element of a group can occur more than once in any row or column of the table, the ▧-box in T can only be inhabited by e, b or c, with each of those possibilities displayed as E, B, and C in Figure 2.2. Repeatedly reemploying Theorem 2.2, we observe that while E leads to two possible group tables, both B and C can only be completed in one way (see Figure 2.3)

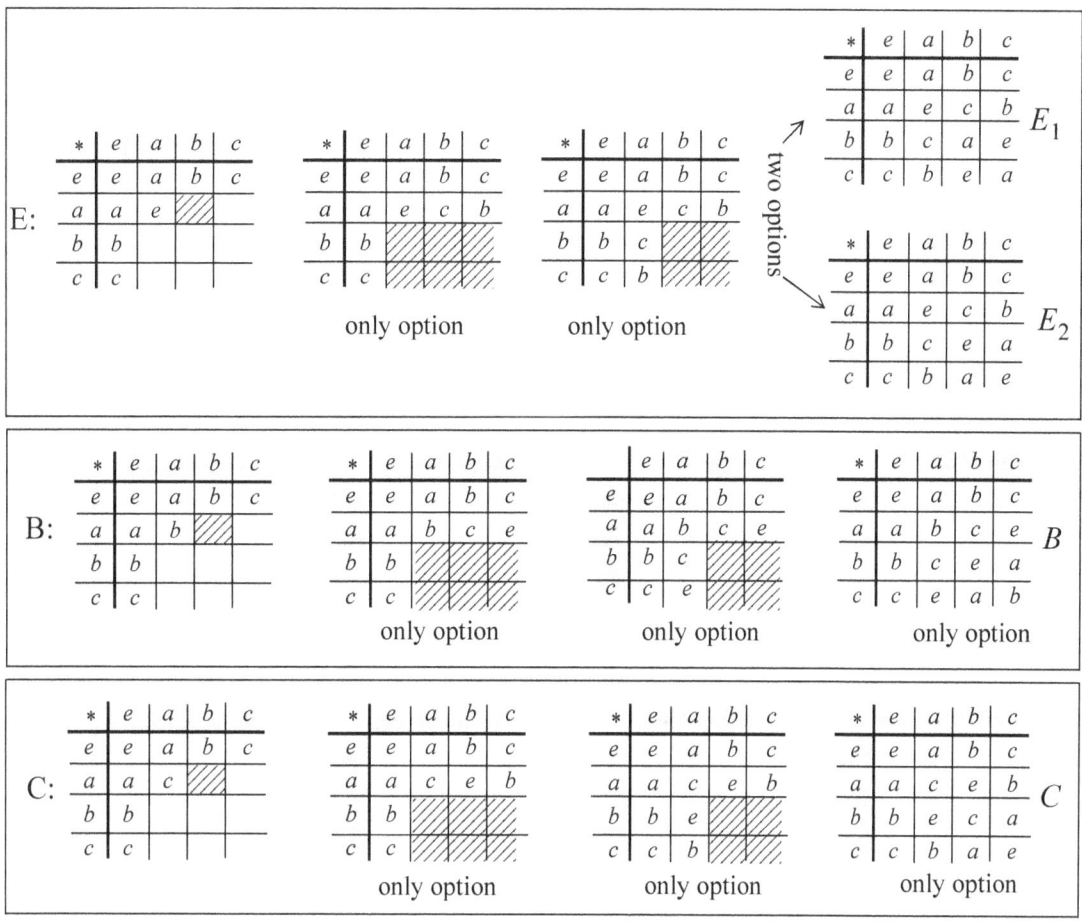

Figure 2.3

At this point we know that there can be at most four groups of order 4, and their corresponding group tables appear in Figure 2.4. The group tables for Z_4 and K of Figure 2.1 are also displayed at the bottom Figure 2.4.

Figure 2.4

While table E_2 and the Klein group table K are identical, those of the remaining four tables in Figure 2.4 look different.

But looks can be deceiving:

To show, for example, that Z_4 and E_1 only differ superficially, we begin by reordering the elements in the first row and first column of Z_4 in Figure 2.5(a) from "0, 1, 2, 3" to "0, 2, 1, 3" [see Figure 2.5 (b)]. We then transform Figure 2.5(b) to E_1 in (c) by replacing the symbols "0, 2, 1, 3" with the symbols "e, a, b, c," respectively, and the operator symbol "$+_4$" with "$*$".

Z_4:

$+_4$	0	1	2	3
0	0	1	2	3
1	1	2	3	0
2	2	3	0	1
3	3	0	1	2

$+_4$	0	2	1	3
0	0	2	1	3
2	2	0	3	1
1	1	3	2	0
3	3	1	0	2

E_1:

$*$	e	a	b	c
e	e	a	b	c
a	a	e	c	b
b	b	c	a	e
c	c	b	e	a

(a)   (b)   (c)

Figure 2.5

This "appearances aside" concept is formalized in Section 4.

So, appearances aside, the group structure of E_1 coincides with that of Z_4. In a similar fashion you can verify that tables B and C of Figure 2.4 only differ from table Z_4 syntactically.

PERMUTATIONS AND SYMMETRIC GROUPS

For any non-empty set X, let $S_X = \{f: X \to X | f \text{ is a bijection}\}$. We then have:

The composition operator "$\circ$" is defined on page 3.

THEOREM 2.4 For any non-empty set X, $\langle S_X, \circ \rangle$ is a group.

PROOF: Turning to Definition 2.1:

Operator. $\forall f, g \in S_X: g \circ f \in S_X$ [Theorem 1.2(c), page 7].

Associative. $\forall f, g, h \in S_X: h \circ (g \circ f) = (h \circ g) \circ f$ [Exercise 51, page 10].

Identity. $\forall f \in S_X: f \circ I_X = I_X \circ f = f$, where $I_X: X \to X$ is the identity function: $I_X(x) = x$ for every $x \in X$.

Inverse. $\forall f \in S_X: f \circ f^{-1} = I_X$ [Theorem 1.1(b), page 5].

> The elements (functions) in S_X are said to be **permutations** (on X), and $\langle S_X, \circ \rangle$ is said to be the **symmetric group** on X.

IN PARTICULAR:

> For $X = \{1, 2, ..., n\}$, $\langle S_X, \circ \rangle$ is called the **symmetric group of degree n**, and will be denoted by S_n.

Let's get our feet wet by considering the symmetric group S_3, the set of permutations on $X = \{1, 2, 3\}$. Since there are $n!$ ways of ordering n objects (Exercise 31, page 20), the group S_3 consists of $3! = 1 \cdot 2 \cdot 3 = 6$ elements:

e	α_1	α_2	α_3	α_4	α_5
$1 \to 1$	$1 \to 2$	$1 \to 3$	$1 \to 1$	$1 \to 3$	$1 \to 2$
$2 \to 2$	$2 \to 3$	$2 \to 1$	$2 \to 3$	$2 \to 2$	$2 \to 1$
$3 \to 3$	$3 \to 1$	$3 \to 2$	$3 \to 2$	$3 \to 1$	$3 \to 3$

Directly below each elements of the first row appears its image under the permutations. The fact that 3 lies below 1 in α_4, for example, simply indicates that the permutation α_4 maps 1 to 3: $1 \to 3$.

In a more compact (and more standard) form (see margin), we write:

$$e = \begin{pmatrix} 1 & 2 & 3 \\ 1 & 2 & 3 \end{pmatrix}, \ \alpha_1 = \begin{pmatrix} 1 & 2 & 3 \\ 2 & 3 & 1 \end{pmatrix}, \ \alpha_2 = \begin{pmatrix} 1 & 2 & 3 \\ 3 & 1 & 2 \end{pmatrix}$$

$$\alpha_3 = \begin{pmatrix} 1 & 2 & 3 \\ 1 & 3 & 2 \end{pmatrix}, \ \alpha_4 = \begin{pmatrix} 1 & 2 & 3 \\ 3 & 2 & 1 \end{pmatrix}, \ \alpha_5 = \begin{pmatrix} 1 & 2 & 3 \\ 2 & 1 & 3 \end{pmatrix}$$

Note that α_0 is the identity function e: $\alpha_0(1) = 1, \alpha_0(2) = 2,$ and $\alpha_0(3) = 3$

The symmetric group S_3
Figure 2.6

Generalizing the above observation we have:

THEOREM 2.5 The symmetric group S_n of degree n contains $n!$ elements.

EXAMPLE 2.1 Referring to the group S_3 featured in Figure 2.6, Determine:

(a) $\alpha_2 \circ \alpha_4$ (b) $\alpha_4 \circ \alpha_2$ (c) $(\alpha_2)^{-1}$

SOLUTION: (a) To find $\alpha_2 \circ \alpha_4$ we first perform α_4 and then apply α_2 to the resulting function values:

From Figure 2.6:

α_2 α_4
$1 \to 3$ $1 \to 3$
$2 \to 1$ $2 \to 2$
$3 \to 2$ $3 \to 1$

$\alpha_4 \quad \alpha_2$
$1 \to 3 \to \mathbf{2}$
$2 \to 2 \to \mathbf{1}$
$3 \to 1 \to \mathbf{3}$

$\Rightarrow \alpha_2 \circ \alpha_4 = \begin{matrix} 1 \to 2 \\ 2 \to 1 \\ 3 \to 3 \end{matrix} = \alpha_5$

(b) Using the standard form we show that $\alpha_4 \circ \alpha_2 = \alpha_3$:

$$\begin{pmatrix} 1 & 2 & 3 \\ 3 & 1 & 2 \\ 1 & 3 & 2 \end{pmatrix} \begin{matrix} \text{first } \sigma_2: \begin{pmatrix} 1 & 2 & 3 \\ 3 & 1 & 2 \end{pmatrix} \\ \text{then } \sigma_4: \begin{pmatrix} 1 & 2 & 3 \\ 3 & 2 & 1 \end{pmatrix} \end{matrix} \Rightarrow \alpha_4 \circ \alpha_2 = \begin{pmatrix} 1 & 2 & 3 \\ 1 & 3 & 2 \end{pmatrix} = \alpha_3$$

(c) Tor arrive at the inverse of the permutation $\alpha_2 = \begin{pmatrix} 1 & 2 & 3 \\ 3 & 1 & 2 \end{pmatrix}$, simply reverse its action:

$$\alpha_2^{-1} = \begin{pmatrix} 1 & 2 & 3 \\ 3 & 1 & 2 \end{pmatrix}^{-1} = \begin{pmatrix} 3 & 1 & 2 \\ 1 & 2 & 3 \end{pmatrix} = \begin{pmatrix} 1 & 2 & 3 \\ 2 & 3 & 1 \end{pmatrix} = \alpha_1$$

Answer:

$$\tau\circ\sigma: \begin{pmatrix} 1 & 2 & 3 & 4 & 5 \\ 5 & 4 & 3 & 2 & 1 \end{pmatrix}$$

$$\sigma\circ\tau: \begin{pmatrix} 1 & 2 & 3 & 4 & 5 \\ 4 & 3 & 2 & 5 & 1 \end{pmatrix}$$

CHECK YOUR UNDERSTANDING 2.4

With reference to the symmetric group S_5, determine $\tau\circ\sigma$ and $\tau\circ\sigma$, where:

$$\sigma = \begin{pmatrix} 1 & 2 & 3 & 4 & 5 \\ 1 & 5 & 2 & 3 & 4 \end{pmatrix} \text{ and } \tau = \begin{pmatrix} 1 & 2 & 3 & 4 & 5 \\ 5 & 3 & 2 & 1 & 4 \end{pmatrix}$$

Adhering to convention, we will start using ab (rather than $a*b$) to denote the binary operation in a generic group. Under this notation, the symbol a^{-1} (rather than a') is used to denote the inverse of a, while e continues to represent the identity element. In a generic abelian group, however, the symbol "+" is typically used to represent the binary operator, with 0 denoting the identity element, and $-a$ denoting the inverse of a. To summarize:

In Summery:

Original Form	Product Form	Sum Form (Reserved **only** for abelian groups)
1. $a*b \in G \ \forall a, b \in G$	1. $ab \in G$	1. $a+b \in G$
2. $a*(b*c) = (a*b)*c$	2. $a(bc) = (ab)c$	2. $a+(b+c) = (a+b)+c$
3. $a*e = a$	3. $ae = a$	3. $a+0 = a$
4. $a*a' = e$	4. $aa^{-1} = e$	4. $a+(-a) = 0$

SOME ADDITIONAL NOTATION:

Referring to the product form, do not express a^{-n} in the form ~~$\frac{1}{a^n}$~~ (there is no "division" in the group).

From its very definition we find that the following exponent rules hold in any group G:
For any $n, m \in Z$:
$a^n a^m = a^{n+m}$
$(a^n)^m = a^{nm}$

In the sum form, it is acceptable utilize the notation $a-b$. By definition:
$a-b = a+(-b)$.

For any positive integer n:
a^n represents $\underbrace{aaa\cdots a}_{n\ a\text{'s}}$
and $a^{-n} = (a^{-1})^n$
We also define a^0 to be e.

For any positive integer n:
na represents $\underbrace{a+a+a+\cdots+a}_{n\ a\text{'s}}$
and $(-n)a = n(-a)$
We also define $0a$ to be 0.

Utilizing the above notation:

DEFINITION 2.4
CYCLIC GROUP

(**Product form**) A group G is **cyclic** if there exists $a \in G$ such that $G = \{a^n | n \in Z\}$.

(**Sum form**) An abelian group G is **cyclic** if there exists $a \in G$ such that $G = \{na | n \in Z\}$.

GENERATOR

In either case we say that the element a is a **generator** of G, and write $G = \langle a \rangle$.

EXAMPLE 2.2 Show that:
(a) Z_6 is cyclic   (b) S_3 is not cyclic.

SOLUTION: (a) Clearly $Z_6 = \langle 1 \rangle$. In fact, as we now show, **5** is also a generator of Z_6 (don't forget that we are summing modulo 6):

	Note that:
$1(5) = 5$	$5 = 0 \cdot 5 + 5$
$2(5) = 5 +_6 5 = 4$	$10 = 1 \cdot 6 + 4$
$3(5) = 5 +_6 5 +_6 5 = 3$	$15 = 2 \cdot 6 + 3$
$4(5) = 5 +_6 5 +_6 5 +_6 5 = 2$	$20 = 3 \cdot 6 + 2$
$5(5) = 5 +_6 5 +_6 5 +_6 5 +_6 5 = 1$	$25 = 4 \cdot 6 + 1$
$6(5) = 5 +_6 5 +_6 5 +_6 5 +_6 5 +_6 5 = 0$	$30 = 5 \cdot 6 + 0$

Since every element of $Z_6 = \{0, 1, 2, 3, 4, 5\}$ is a multiple of **5**, we conclude that $Z_6 = \langle 5 \rangle$.

(b) We could use a brute-force method to verify, directly, that no element of S_3 generates all of S_3. Instead, we appeal to the following theorem [and Example 2.2(b)] to draw the desired conclusion.

THEOREM 2.6 *Every cyclic group is abelian.*

PROOF: Let $G = \langle a \rangle = \{a^n | n \in Z\}$. For any two elements a^s and a^t in G (not necessarily distinct) we have:

$$a^s a^t = a^{s+t} = a^{t+s} = a^t a^s$$

CHECK YOUR UNDERSTANDING 2.5

(a) Show that 1 and 5 are the only generators of Z_6.

(b) Show that S_2 is cyclic.

(c) Show that S_n is not cyclic for any $n > 2$.

Answer: See page A-8.

At this point we have two groups of order 6 at our disposal:

$$\langle Z_6, +_n \rangle \text{ and } S_3$$

Do these groups differ only superficially, or are they really different in some algebraic sense? They do differ algebraically in that one is cyclic while the other is not, and also in that one is abelian while the other is not.

EXERCISES

Exercise 1-11. Determine if the given set is a group under the given operator. If not, specify why not. If it is, indicate whether or not the group is abelian, and whether or not it is cyclic. If it is cyclic, find a generator for the group.

1. The set $\{2n | n \in Z\}$ of even integers under addition.

2. The set $\{2n + 1 | n \in Z\}$ of odd integers under addition.

3. The set of integers Z, with $a*b = c$, where c is the smaller of the two integers a and b (the common value if $a = b$).

4. The set Q^+ of positive rational numbers, with $a*b = \dfrac{ab}{2}$.

5. The set $\{x \in \Re | x \neq 0\}$, with $a*b = \dfrac{a^2}{b}$.

6. The set $\{0, 2, 4, 6, 8\}$ under the operation of addition modulo 10.

7. The set $\{0, 1, 2, 3\}$ under multiplication modulo 4. (For example: $2*3 = \mathbf{2}$, since $2 \cdot 3 = 6 = 1 \cdot 4 + \mathbf{2}$; and $3*3 = \mathbf{1}$, since $3 \cdot 3 = 9 = 2 \cdot 4 + \mathbf{1}$.)

8. The set $\{0, 1, 2, 3, 4\}$ under multiplication modulo 5. (See Exercise 7.)

9. The set $\{a + b\sqrt{2} | a, b \in Z\}$ under addition.

10. The set $\{a + b\sqrt{2} | a, b \in Q$ with not both a and b equal to $0\}$ under the usual multiplication of real numbers.

11. The set $Z \times Z = \{(a, b) | a, b \in Z\}$, with $(a, b) + (c, d) = (a + c, b + d)$.

Exercise 12-23. Referring to the group S_3:

$$e = \begin{pmatrix} 1 & 2 & 3 \\ 1 & 2 & 3 \end{pmatrix}, \alpha_1 = \begin{pmatrix} 1 & 2 & 3 \\ 2 & 3 & 1 \end{pmatrix}, \alpha_2 = \begin{pmatrix} 1 & 2 & 3 \\ 3 & 1 & 2 \end{pmatrix}, \alpha_3 = \begin{pmatrix} 1 & 2 & 3 \\ 1 & 3 & 2 \end{pmatrix}, \alpha_4 = \begin{pmatrix} 1 & 2 & 3 \\ 3 & 2 & 1 \end{pmatrix}, \alpha_5 = \begin{pmatrix} 1 & 2 & 3 \\ 2 & 1 & 3 \end{pmatrix}$$

determine:

12. $\alpha_2 \alpha_4$ and $\alpha_4 \alpha_2$

13. α_3^2 and α_3^3

14. α_3^n for $n \in Z^+$.

15. α_1^n for $n \in Z^+$.

16. α_3^{-2} and α_3^{-3}

17. α_3^{-n} for $n \in Z^+$.

18. α_3^{-n} for $n \in Z^+$.

19. α_2^2 and α_2^3

20. α_2^n for $n \in Z^+$.

21. α_2^n for $n \in Z^+$.

22. α_2^{-2} and α_2^{-3}

23. α_2^{-n} for $n \in Z^+$.

Exercise 24-33. For

$$\alpha = \begin{pmatrix} 1 & 2 & 3 & 4 & 5 & 6 \\ 2 & 3 & 4 & 5 & 6 & 1 \end{pmatrix} \qquad \beta = \begin{pmatrix} 1 & 2 & 3 & 4 & 5 & 6 \\ 2 & 1 & 4 & 3 & 6 & 5 \end{pmatrix} \qquad \gamma = \begin{pmatrix} 1 & 2 & 3 & 4 & 5 & 6 \\ 6 & 5 & 4 & 3 & 2 & 1 \end{pmatrix}$$

Determine:

24. $\alpha\beta$ 25. $\beta\alpha$ 26. $\beta\gamma$ 27. $\gamma\beta$ 28. $\alpha\beta\gamma$

29. α^5 30. α^{100} 31. α^{101} 32. β^{100} 33. β^{101}

34. Let $S = \{1\}$. Show that $\langle S, * \rangle$ with $1 * 1 = 1$ is a group. Is the group abelian? Cyclic?

35. Is $\langle M_{2 \times 2}, + \rangle$ with $\begin{bmatrix} a & b \\ c & d \end{bmatrix} + \begin{bmatrix} \bar{a} & \bar{b} \\ \bar{c} & \bar{d} \end{bmatrix} = \begin{bmatrix} a+\bar{a} & b+\bar{b} \\ c+\bar{c} & d+\bar{d} \end{bmatrix}$ a group? If so, is it abelian? Cyclic?

36. Is $\langle M_{2 \times 2}, * \rangle$ with $\begin{bmatrix} a & b \\ c & d \end{bmatrix} * \begin{bmatrix} \bar{a} & \bar{b} \\ \bar{c} & \bar{d} \end{bmatrix} = \begin{bmatrix} a\bar{a} & b\bar{b} \\ c\bar{c} & d\bar{d} \end{bmatrix}$ a group? If so, is it abelian? Cyclic?

37. Is $\langle M_{2 \times 2}, * \rangle$ with $\begin{bmatrix} a & b \\ c & d \end{bmatrix} * \begin{bmatrix} \bar{a} & \bar{b} \\ \bar{c} & \bar{d} \end{bmatrix} = \begin{bmatrix} a\bar{a}+b\bar{c} & a\bar{b}+b\bar{d} \\ c\bar{a}+d\bar{c} & c\bar{b}+d\bar{d} \end{bmatrix}$ a group? If so, is it abelian? Cyclic?

38. Let $S = \{a, b, c\}$ along with the binary operator:

*	a	b	c
a	a	b	c
b	b	b	c
c	c	c	c

Is $\langle S, * \rangle$ a group?

39. Let $S = \{0, 1, 2\}$ along with the binary operator:

*	2	0	1
2	2	0	1
0	0	1	2
1	1	2	0

Is $\langle S, * \rangle$ a group?

40. Let $S = \{(x, y) | x, y \in \Re\}$. Show that $\langle S, * \rangle$ with $(x, y) * (\bar{x}, \bar{y}) = (x + \bar{x} - 1, y + \bar{y} + 1)$ is a group. Is the group abelian? Cyclic?

41. For $n \geq 0$, let P_n denote the set of polynomials of degree less than or equal to n. Show that

$$\langle P_n, * \rangle \text{ with } \left(\sum_{i=0}^{n} a_i x^i \right) * \left(\sum_{i=0}^{n} b_i x^i \right) = \sum_{i=0}^{n} (a_i + b_i) x^i \text{ is a group. Is the group abelian?}$$

42. Let $S = \{(x, y) | x, y \in \Re\}$. Show that $\langle S, * \rangle$ with $(x, y) * (\bar{x}, \bar{y}) = (x + \bar{x} + 2, y + \bar{y})$ is a group. Is the group abelian?

43. Let Q denote the set of rational numbers. Show that $\langle Q, * \rangle$ with $a*b = a+b+ab$ is not a group.

44. Let $\overline{Q} = \{a \in Q | a \neq -1\}$. Show that $\langle \overline{Q}, * \rangle$ with $a*b = a+b+ab$ is a group. Is the group abelian?

45. Let $G = \{e, a_1, a_2, ..., a_{n-1}\}$. Show that the function $f_{a_i}: G \to G$ given by $f_{a_i}(g) = a_i g$ is a bijection

46. (a) Give an example of a group G in which the exponent law $(ab)^n = a^n b^n$ does not hold in a group G, for $n \in Z^{""}$

    (b) Prove that the exponential law $(ab)^n = a^n b^n$ does hold if the group G is abelian.

    (c) Express the property $(ab)^n = a^n b^n$ in sum-notation form.

47. Let G be a group and $a, b, c \in G$. Show that if $ab = ac$, then $b = c$.

48. Let a be an element in a group G. Show that if $|\langle a \rangle| = 2$, then $ab = ba$ for ever $b \in G$.

49. Let p and q be distinct primes numbers. Find the number of generators of Z_{pq}.

50. (a) Show that the group Z_n of Theorem 2.1 is cyclic for any $n \in Z^+$.

    (b) Prove that $m \in Z_n$ is a generator of Z_n if and only if m and n are relatively prime. Suggestion: Consider Theorem 1.7, page 23.

51. Let $F(\Re)$ denote the set of all real-valued functions. For f and g in $F(\Re)$, let $f+g$ be given by $(f+g)(x) = f(x) + g(x)$. Show that $\langle F(\Re), + \rangle$ is a group. Is the group abelian?

52. Let $G = \{e, a_1, a_2, ..., a_{n-1}\}$ be a group. Show that the function $f_{a_i}(g) = a_i g$ is a bijection.

53. Prove Theorem 2.2.

54. Let G and H be groups. Let $G \times H = \{(g, h) | g \in G, h \in H\}$ with:
$$(g, h) * (\bar{g}, \bar{h}) = (g\bar{g}, h\bar{h})$$

    (a) Show that $\langle G \times H, * \rangle$ is a group.

    (b) Prove that $\langle G \times H, * \rangle$ is abelian if and only both G and H are abelian.

55. Let X be a set and let $P(X)$ be the set of all subsets of X. Is $\langle P(X), * \rangle$ a group if:

    (a) $A * B = A \cup B$ (b) $A * B = A \cap B$

PROVE OR GIVE A COUNTEREXAMPLE

56. The set $\Re$ of real numbers under multiplication is a group

57. The set $\Re^+ = \{r \in \Re | r > 0\}$ of positive real numbers under multiplication.

58. Let G be a group and $a, b, c \in G$. If $b \neq c$, then $ab \neq ac$.

59. The group S_3 contains four elements α such that $\alpha^2 = e$ and three elements β such that $\beta^3 = e$.

60. The group $\langle S_\Re, \circ \rangle$ is abelian.

61. Let G be a group and $a, b \in G$. If $ab = b$, then $ac = c$ for every $c \in G$.

62. Let G be a group and $a, b \in G$. If $ab = ba$, then $ac = ca$ for every $c \in G$.

63. The cyclic group $\langle Z, + \rangle$ has exactly two distinct generators.

§2. ELEMENTARY PROPERTIES OF GROUPS

We begin by recalling the group axioms, featuring both the product and sum notations:

	Product Form	Sum Form (Typically reserved for abelian groups)
Closure	$ab \in G$	$a + b \in G$
Axiom	1. $a(bc) = (ab)c$	1. $a + (b + c) = (a + b) + c$
Identity Axiom	2. $ae = a$	2. $a + 0 = a$
Inverse Axiom	3. $aa^{-1} = e$	3. $a + (-a) = 0$

For aesthetic reasons, a set of axioms should be independent, in that no axiom or part of an axiom is a consequence of the rest. One should not, for example, replace Axiom 2 in Definition 2.1, page 41:

$\exists e \in G \ni ae = a \forall a \in G$

with:

$\exists e \in G \ni ae = ea = a \forall a \in G$

In sum form:
$a + (-a) = 0 \Rightarrow (-a) + a = 0$
$a + 0 = a \Rightarrow 0 + a = a$

Actually, as we show below, both the identity element of Axiom 2 and the inverse elements of Axiom 3 work on both sides; but first:

LEMMA 2.1 Let G be a group. If $a \in G$ is such that $a^2 = a$, then $a = e$.

PROOF:
$aa = a \Rightarrow (aa)a^{-1} = aa^{-1} \underset{\text{Axiom 1}}{\Rightarrow} a(aa^{-1}) = e \underset{\text{Axiom 3}}{\Rightarrow} ae = e \underset{\text{Axiom 2}}{\Rightarrow} a = e$

THEOREM 2.7 Let G be a group. For $a \in G$:

(a) $aa^{-1} = e \Rightarrow a^{-1}a = e$

(b) $ae = a \Rightarrow ea = a$

PROOF:

(a) $(a^{-1}a)(a^{-1}a) \underset{\text{Axiom 1}}{=} [a^{-1}(aa^{-1})]a \underset{\text{Axiom 3}}{=} (a^{-1}e)a \underset{\text{Axiom 2}}{=} a^{-1}a$

We now know that $(a^{-1}a)(a^{-1}a) = a^{-1}a$.

Applying Lemma 2. we then have: $a^{-1}a = e$.

(b) $ea \underset{\text{Axiom 3}}{=} (aa^{-1})a \underset{\text{Axiom 1}}{=} a(a^{-1}a) \underset{\text{part (a)}}{=} ae \underset{\text{Axiom 3}}{=} a$

Axioms 2 and 3 stipulate the existence of an identity and of inverses in a group. Are they necessarily unique? Yes:

THEOREM 2.8 (a) There is but one identity in a group G.

(b) Every element in G has a unique inverse.

2.2 Elementary Properties of Groups

PROOF: (a) We assume that e and $\bar{e}$ are identities, and go on to show that $e = \bar{e}$:

$$e = e\bar{e} = \bar{e}$$

(Since $\bar{e}$ is an identity; Since e is an identity)

(b) We assume that a^{-1} and $\bar{a}^{-1}$ are inverses of $a \in G$, and show that $a^{-1} = \bar{a}^{-1}$:

Since aa^{-1} and $a\bar{a}^{-1}$ are both equal to **the** identity they must be equal to each other:

$$aa^{-1} = a\bar{a}^{-1}$$

Multiply both sides by a^{-1}: $\quad a^{-1}(aa^{-1}) = a^{-1}(a\bar{a}^{-1})$

Associativity: $\quad (a^{-1}a)a^{-1} = (a^{-1}a)\bar{a}^{-1}$

$$ea^{-1} = e\bar{a}^{-1}$$

$$a^{-1} = \bar{a}^{-1}$$

CHECK YOUR UNDERSTANDING 2.6

Show that if a, b, c are elements of a group such that $abc = e$, then $bca = e$.

Answer: See page A-9.

Both the **left and right cancellation laws** hold in groups:

Sum form:
$a + b = c + b \Rightarrow a = c$
$b + a = b + c \Rightarrow a = c$

THEOREM 2.9 In any group G:
(a) If $ab = cb$, then $a = c$.
(b) If $ba = bc$, then $a = c$

PROOF:

Just in case you are asking yourself:
What if b is 0 and has no inverse?
Tisk, every element in a group has an inverse.

(a)
$$ab = cb$$
$$(ab)b^{-1} = (cb)b^{-1}$$
$$a(bb^{-1}) = c(bb^{-1})$$
$$ae = ce$$
$$a = c$$

(b)
$$ba = bc$$
$$b^{-1}(ba) = b^{-1}(bc)$$
$$(b^{-1}b)a = (b^{-1}b)c$$
$$ea = ec$$
$$a = c$$

CHECK YOUR UNDERSTANDING 2.7

PROVE OR GIVE A COUNTEREXAMPLE:
(a) In any group G, if $ab = bc$ then $a = c$.
(b) In any abelian group G, if $a + b = b + c$ then $a = c$.

Answer: See page A-9.

In the real number system, do linear equations $ax = b$ have unique solutions for every $a, b \in \Re$? No: the equation $0x = 5$ has no solution, while the equation $0x = 0$ has infinitely many solutions. This observation assures us once more that the reals is not a group under multiplication, since:

THEOREM 2.10 Let G be a group. For any $a, b \in G$, the linear equations $ax = b$ and $ya = b$ have unique solutions in G.

PROOF: Existence:

(a)
$$ax = b$$
$$a^{-1}(ax) = a^{-1}b$$
$$(a^{-1}a)x = a^{-1}b$$
$$ex = a^{-1}b$$
$$x = a^{-1}b$$

(b)
$$ya = b$$
$$(ya)a^{-1} = ba^{-1}$$
$$y(aa^{-1}) = ba^{-1}$$
$$ye = ba^{-1}$$
$$y = ba^{-1}$$

Uniqueness: We assume (as usual) that there are two solutions, and then proceed to show that they are equal:

(a) $ax_1 = b$ and $ax_2 = b \Rightarrow ax_1 = ax_2 \overset{\text{Theorem 2.9(b)}}{\Rightarrow} x_1 = x_2$

(b) $y_1 a = b$ and $y_2 a = b \Rightarrow y_1 a = y_2 a \overset{\text{Theorem 2.9(a)}}{\Rightarrow} y_1 = y_2$

CHECK YOUR UNDERSTANDING 2.8

Since the set of real numbers under addition is a group, 2.9 applies. Show, **directly**, that any linear equation in $\langle \Re, + \rangle$ has a unique solution.

Answer: See page A-9.

The inverse of a product is the product of the inverses, but in reverse order:

This is another *shoe-sock theorem* (see page 8).

THEOREM 2.11 For every a, b in a group G:
$$(ab)^{-1} = b^{-1}a^{-1}$$

PROOF: To show that $b^{-1}a^{-1}$ is the inverse of ab is to show that $(b^{-1}a^{-1})(ab) = e$. No problem:

$$(b^{-1}a^{-1})(ab) = b^{-1}(a^{-1}a)b = b^{-1}eb = b^{-1}b = e$$

2.2 Elementary Properties of Groups 57

> **CHECK YOUR UNDERSTANDING 2.9**
>
> Give an example of a group G for which $(ab)^{-1} = a^{-1}b^{-1}$ does not hold for every $a, b \in G$.

Answer: See page A-9.

The axiom of a group G assures us that an expression such as abc, sans parentheses, is unambiguous [since $(ab)c$ and $a(bc)$ yield the same result]. It is plausible to expect that this nicety extends to any product $a_1 a_2 \cdots a_n$ of elements of G. Plausible, to be sure; but more importantly, True:

THEOREM 2.12 Let $a_1 a_2 \cdots a_n \in G$. The product expression $a_1 a_2 \cdots a_n$ is unambiguous in that its value is independent of the order in which adjacent factors are multiplied.

PROOF: [By induction (page 13)]:

I. The claim holds for $n = 3$ (the axiom).

II. Assume the claim holds for $n = k$, with $k > 3$.

III. (Now for the fun part) We show the claim holds for $n = k + 1$:

Let x denote the product $a_1 a_2 \cdots a_{k+1}$ under a certain pairing of its elements, and y the product under another pairing of its elements. We are to show that $x = y$. Let's do it:

Assume that one starts the two multiplication processes with the following pairing for x and y:

$$x = \overset{A}{(a_1 a_2 \cdots a_i)} \overset{B}{(a_{i+1} \cdots a_{k+1})} \text{ and } y = \overset{C}{(a_1 a_2 \cdots a_j)} \overset{D}{(a_{j+1} \cdots a_{k+1})}$$

Case 1. $i = j$: By the induction hypothesis (II), no matter how the products in A and C are performed, A will equal C. The same can be said concerning B and D. Consequently $x = AB = CD = y$.

Case 2. Assume, without loss of generality, that $i < j$. Breaking the "longer" product B into two pieces M and D we have:

$$x = \overset{A}{(a_1 a_2 \cdots a_i)} \overset{M}{(a_{i+1} \cdots a_j)} \overset{D}{(a_{j+1} \cdots a_{k+1})}$$

By the induction hypothesis, A, M, and D are well defined (independent of the pairing of its elements in their products). Bringing us to:

$$x = AB = A(MD) \overset{\downarrow}{=} (AM)D = CD = y$$

I: Claim holds for $n = 3$

CHECK YOUR UNDERSTANDING 2.10

Use the Principle of Mathematical Induction, to show that for any $a_1 a_2 \cdots a_n \in G$:

$$(a_n \cdots a_2 a_1)^{-1} = a_1^{-1} a_2^{-1} \cdots a_n^{-1}$$

Answer: See page A-9.

THEOREM 2.13 For any given element a of a finite group G:

$$a^m = e \text{ for some } m \in Z^+.$$

PROOF: Let G be of order n. Surely not all of the $n+1$ elements $a, a^2, a^3, \ldots, a^{n+1}$ can be distinct. Choose $1 \leq s < t \leq n+1$ such that $a^t = a^s$. Since $a^t a^{-s} = a^{t-s} = e$:

$$a^m = e, \text{ for } m = t - s.$$

DEFINITION 2.5

ORDER OF AN ELEMENT OF G

Let G be a group, and let $a \in G$ be such that $a^m = e$ for $m \in Z^+$. The smallest such m is called the **order of a** and is denoted by $o(a)$. If no such m exists, then a is said to have infinite order.

In the additive notation, $a^m = e$ translates to $na = 0$; which is to say:
$a + a + \ldots + a = 0$
(sum of n a's)

EXAMPLE 2.3 (a) Determine the order of the element 4 in the group $\langle Z_6, +_6 \rangle$.

(b) Determine the order of the element

$$\sigma = \begin{pmatrix} 1 & 2 & 3 & 4 & 5 \\ 3 & 2 & 4 & 1 & 5 \end{pmatrix}$$

in the symmetric group S_5.

SOLUTION: (a) Since:

$$1(4) = 4$$
$$2(4) = 4 +_6 4 = 2$$
$$3(4) = 4 +_6 4 +_6 4 = 2 +_6 4 = 0$$

The element 4 has order 3 in Z_6.

(b) Since:

$$\begin{matrix} & \begin{pmatrix} 1 & 2 & 3 & 4 & 5 \\ \sigma & 3 & 2 & 4 & 1 & 5 \\ \sigma^2 & 4 & 2 & 1 & 3 & 5 \\ \sigma^3 & 1 & 2 & 3 & 4 & 5 \end{pmatrix} & e \end{matrix}$$

The element $\sigma = \begin{pmatrix} 1 & 2 & 3 & 4 & 5 \\ 3 & 2 & 4 & 1 & 5 \end{pmatrix}$ has order 3 in S_5.

CHECK YOUR UNDERSTANDING 2.11

(a) Determine the order of the element $\sigma = \begin{pmatrix} 1 & 2 & 3 & 4 \\ 2 & 3 & 4 & 1 \end{pmatrix}$ in S_4.

(b) Determine the order of the element 4 in Z_{24}.

(c) Let $a \in Z_n$. Prove that $o(m) = \dfrac{n}{gcd(a, n)}$
↑
See Definition 1.8, page 22

Answer: (a) 4 (b) 6
(c) See page A-10.

Note: There is no "subtraction" in a group $\langle G, + \rangle$. For convenience, however, for given $a, b \in G$, we define the symbol $a - b$ as follows:
$$a - b = a + (-b)$$
(add the additive inverse of b to a)

There is no "division" in a group $\langle G, \cdot \rangle$. In this setting, however, one does not ever substitute the symbol $\dfrac{a}{b}$ for ab^{-1}. Why not? Convention.

EXERCISES

1. Let G be a group and $a, b, c \in G$. Solve for x, if:
   (a) $axa^{-1} = e$ (b) $axa^{-1} = a$ (c) $axb = c$ (d) $ba^{-1}xab^{-1} = ba$

2. Let G be a group. Prove that $(a^{-1})^{-1} = a$ for every $a \in G$.

3. Prove that for any element a in a group G the functions $f_a: G \to G$ given by $f_a(b) = ab$ and the function $g_a: G \to G$ given by $g_a(b) = ba$ are bijections.

4. Let a be an element of a group G. Show that $G = \{ab | b \in G\}$

5. Let G be a group and let $a \in G$. Show that if there exists one element $x \in G$ for which $ax = x$, then $a = e$.

6. Let a be an element of a group G for which there exists $b \in G$ such that $ab = b$. Prove that $a = e$.

7. Prove that a group G is abelian if and only if $(ab)^{-1} = a^{-1}b^{-1}$ for every $a, b \in G$.

8. Let G be group for which $a^{-1} = a$ for every $a \in G$. Prove that G is abelian.

9. Let G be group for which $(ab)^2 = a^2b^2$ for every $a, b \in G$. Prove that G is abelian.

10. Let G be a finite group consisting of an even number of elements. Show that there exists $a \in G$, $a \neq e$, such that $a^2 = e$.

11. (a) Let G be a group. Show that if, for any $a, b \in G$, there exist three consecutive integers i such that $(ab)^i = a^i b^i$ then G is abelian.
    (b) Let G be an abelian group. Show that for any $a \in G$ and $n \in Z$: $(-n)a = n(-a)$.

12. Let $*$ be an associative operator on a set S. Assume that for any $a, b \in S$ there exists $c \in S$ such that $a*c = b$, and an element $d \in S$ such that $d*a = b$. Show that $\langle S, * \rangle$ is a group.

13. Let G be a group and $a \in G$. Define a new operation $*$ on G by $b*c = ba^{-1}c$ for all $b, c \in G$. show that $\langle G, * \rangle$ is a group.

14. Let G be a group and $a, b \in G$. Use the Principle of Mathematical Induction to show that for any positive integer n: $(a^{-1}ba)^n = a^{-1}b^n a$.

15. Let a and b be elements of a group G. Show that if ab has finite order n, then ba also has order n.

16. Let G be a cyclic group of order n. Show that if m is a positive integer, then G has an element of order m if and only if m divides n.

17. Let G be a group. Show that for every element $a \in G$ and for any $n \in Z$: $a^{-n} = (a^{-1})^n$.

18. Let G be a finite group, and $a, b \in G$. Prove that the elements a, a^{-1} and bab^{-1} have the same order.

19. List the order of each element in the Symmetric group S_3 of Figure 2.6, page 47.

20. Let $a \in G$ be of order n. Prove that $a^s = a^t$ if and only if n divides $s - t$.

21. Prove that if $a^2 = e$ for every element a in a group G, then G is abelian.

22. Let $*$ be an associative operator on a finite set S. Show that if both the left and right cancellation laws of Theorem 2.8 hold under $*$, then $\langle S, * \rangle$ is a group.

	PROVE OR GIVE A COUNTEREXAMPLE	

23. If a, b, c are elements of a group such that $abc = e$, then $cba = e$.

24. In any group G there exists exactly one element a such that $a^2 = a$.

25. In any group G, $(ab)^{-2} = b^{-2}a^{-2}$.

26. Let G be a group. If $abc = bac$ then $ab = ba$.

27. Let G be a group. If $abcd = bacd$ then $ab = ba$.

28. Let G be a group. If $(abc)^{-1} = a^{-1}b^{-1}c^{-1}$ then $a = c$.

§3. SUBGROUPS

DEFINITION 2.6
SUBGROUP
A **subgroup** of a group G is a nonempty subset H of G which is itself a group under the imposed binary operation of G.

As it turns out, apart from closure, to determine whether or not a nonempty subset of a group is a subgroup you need but challenge Axiom 3:

GROUP AXIOMS
Closure: $ab \in G$
Axiom 1. $a(bc) = (ab)c$
Axiom 2. $ae = a$
Axiom 3. $aa^{-1} = e$

THEOREM 2.14 A nonempty subset S of a group G is a subgroup of G if and only if:

(i) S is closed with respect to the operation in G.

(ii) $s \in S$ implies that $s^{-1} \in S$.

PROOF: If S is a subgroup, then (i) and (ii) must certainly be satisfied.

Conversely, if (i) and (ii) hold in S, then Axioms 1 and 2 also hold:

Axiom 1: Since $a(bc) = (ab)c$ holds for every $a, b, c \in G$, that associative property must surely hold for every $a, b, c \in S$.

Axiom 2: Since $ae = a$ for every $a \in G$, then surely $se = s$ for every $s \in S$. It remains to be shown that $e \in S$. Lets do it:

Choose any $s \in S$. By (ii): $s^{-1} \in S$.

By (i): $ss^{-1} = e \in S$.

When challenging if $S \subset G$ is a subgroup, we suggest that you first determine if it contains the identity element. For if not, then S is not a subgroup, period. If it does, then $S \neq \emptyset$ and you can then proceed to challenge (i) and (ii) of Theorem 2.14.

For example:

$5Z = \{..., -10, -5, 0, 5, 10, ...\}$

EXAMPLE 2.4 Show that for any fixed $n \in Z$ the subset
$$nZ = \{nm | m \in Z\}$$
is a subgroup of $\langle Z, + \rangle$.

SOLUTION: Since $0 = n0 \in nZ$, $nZ \neq \emptyset$.

(i) nZ is closed under addition:
$$nm_1 + nm_2 = n(m_1 + m_2) \in nZ$$

We remind you that, under addition, $-a$ rather than a^{-1} is used to denote the inverse of a.

(ii) For any $nm \in nZ$:
$$-(nm) = n(-m) \in nZ$$

Conclusion: nZ is a subgroup of Z (Theorem 2.14).

CHECK YOUR UNDERSTANDING 2.12

Answer: See page A-10.

The previous example assures us that $3Z$ is a subgroup of $\langle Z, + \rangle$. As such, it is itself a group. Show that $6Z$ is a subgroup $3Z$.

You are invited to show in the exercises that the following result holds for any collection of subgroups of a given group:

THEOREM 2.15 If H and K are subgroups of a group G, then $H \cap K$ is also a subgroup of G.

PROOF: Since H and K are subgroups, each contains the identity element. It follows that $e \in H \cap K$ and that therefore $H \cap K \neq \emptyset$. We now verify that conditions (i) and (ii) of Theorem 2.13 are satisfied:

(i) (Closure) If $a, b \in H \cap K$, then $a, b \in H$ and $a, b \in K$. Since H and K are subgroups, $ab \in H$ and $ab \in K$. It follows that $ab \in H \cap K$.

(ii) (Inverses) If $a \in H \cap K$, then $a \in H$ and $a \in K$. Since H and K are subgroups, $a^{-1} \in H$ and $a^{-1} \in K$. consequently, $a^{-1} \in H \cap K$.

CHECK YOUR UNDERSTANDING 2.13

Answer: See page A-10.

Let H and K be subgroups of a G for which $H \cap K = \{e\}$. Prove:
$$h_1 k_1 = h_2 k_2 \Rightarrow h_1 = h_2 \text{ and } k_1 = k_2.$$

We recall the definition of a cyclic group appearing on page 48:

> A group G is **cyclic** if there exists $a \in G$ such that $G = \{a^n | n \in Z\}$.

DEFINITION 2.7 Let G be a group, and $a \in G$. The cyclic group $\langle a \rangle = \{a^n | n \in Z\}$ is called the cyclic **subgroup** of G **generated** by a.
(In sum form: $\langle a \rangle = \{na | n \in Z\}$)

CHECK YOUR UNDERSTANDING 2.14

Answer: $\langle 3 \rangle = Z_8$
$\langle 4 \rangle = \{0, 4\}$

For $G = Z_8$, determine $\langle 3 \rangle$ and $\langle 4 \rangle$. (Use sum notation.)

THEOREM 2.16   Every subgroup of a cyclic group is cyclic.

PROOF: Let H be a subgroup of $G = \langle a \rangle$. If $a = e$, then $H = \{e\}$, which is cyclic. If $a \neq e$ then let m be the smallest positive integer such that $a^m \in H$. We show $H = \langle a^m \rangle$ by showing that every $a^n \in H$ is a power of a^m:

Employing the Division Algorithm of page 21, we chose integers q and r, with $0 \leq r < m$, such that: $n = mq + r$. And so we have:

$$a^n = a^{mq+r} = (a^m)^q a^r \quad (*) \text{ or: } a^r = (a^m)^{-q} a^n \quad (**)$$

Since a^n and a^m are both in H, and since H is a group: $(a^m)^{-q} a^n \in H$. Consequently, from (**): $a^r \in H$.

Since $0 \leq r < m$ and since m is the smallest positive integer such that $a^m \in H$: $r = 0$. Consequently, from (*):

$$a^n = (a^m)^q a^0 = (a^m)^q \text{ — a power of } a^m.$$

CHECK YOUR UNDERSTANDING 2.15

Let $G = \langle a \rangle$ with $|G| = n$. Let $b \in G$ with $b = a^s$. Prove that:

$$o\langle b \rangle = \frac{n}{gcd(n,s)}$$

Answer: See page A-11.

SUBGROUPS GENERATED BY SUBSETS OF A GROUP

We have seen that any element a in a group G can be used to generate a subgroup of G — namely the cyclic group generated by a:

$$\langle a \rangle = \{a^n | n \in Z\}$$

Generalizing the above concept, we start off with a nonempty subset A of G, and consider the set $\langle A \rangle$ of all elements of G consisting of finite products of elements of $\langle A \rangle$, wherein repetitions of its elements may occur. For example, if $A = \{a, b, c\}$, then:

$$a^3, c^{-2}b^3, aa^{-1} = e, \text{ and } ab^2c^{-3}a^3c^2a \text{ are all in } \langle A \rangle.$$

In the event that G is abelian, the elements of $\langle A \rangle$ can be expressed in a non-repetition form, as with:

$ab^2c^{-3}a^3c^2a = a^5b^2c^{-1}$

Note that, by its very definition, $\langle A \rangle$ is a subgroup of G:

$\langle A \rangle$ is certainly not empty and closed under multiplication.

Moreover, the inverse of any element in $\langle A \rangle$ is again of the form which positions it in $\langle A \rangle$. For example:

$$(ab^2c^{-3}a^3c^2a)^{-1} = a^{-1}c^{-2}a^{-3}c^3b^{-2}a^{-1}$$

(see CYU 2.10, page 58)

Bringing us to:

DEFINITION 2.8
GENERATED SUBGROUP
Let A be a nonempty subset of a group G. The **subgroup of G generated by A**, denoted by $\langle A \rangle$, consisting of all finite products of elements of $\langle A \rangle$

In particular, here is the subgroup of $\langle Z, + \rangle$ generated by $\{2, 12\}$:

$$\langle 2, 12 \rangle = \{2^n 12^m | n, m \in Z\} \underset{\uparrow}{=} \{2^n 3^m | n, m \in Z\}$$
$$\text{since } 12 = 2^2 \cdot 3$$

Note that commutativity enables us to gather all of the 2's and 3's together.

THEOREM 2.17 Let A be a nonempty subset of a group G. The following are equivalent:

(i) $S = \langle A \rangle$

(ii) S is the intersection of all subgroups of G containing A.

PROOF: $(i) \Rightarrow (ii)$: Since subgroups are closed under multiplication, any subgroup of G that contains A, including the subgroup $\langle A \rangle$ has to contain $\langle A \rangle$. It follows that $\langle A \rangle$ is the intersection of all subgroups of G that contain A.

$(ii) \Rightarrow (i)$ Your turn:

CHECK YOUR UNDERSTANDING 2.16

Verify that $(ii) \Rightarrow (i)$.

Answer: See page A-11.

Here is a particularly important result:

THEOREM 2.18
(Lagrange)
If G is a finite group and H is a subgroup of G, then the order of H divides the order of G:

$$|H| \big| |G|$$

(see Definition 2.2, page 43)

Joseph-Louis Lagrange (1736-1813).

To illustrate: If a group G contains 35 elements, it cannot contain a subgroup of 8 elements, as 8 does not divide 35.

We will eventually show that the converse of Lagrange's Theorem holds for abelian groups. It does not, however, hold in general (Exercise 27, page 102).

A proof of Lagrange's Theorem is offered at the end of the section. At this point, we turn to a few of its consequences, beginning with:

THEOREM 2.19 Any group G of prime order is cyclic.

PROOF: Let $|G| = p$, where p is prime. Since $p \geq 2$, we can choose an element $a \in G$ distinct from e. By Lagrange's theorem, the order of the cyclic group $\langle a \rangle = \{a^n | n \in Z\}$ must divide p. But only 1 and p

divide p, and since $\langle a \rangle$ contains more than one element, it must contain p elements, and is therefore all of G.

The symmetric group S_3 is an example of a non-abelian group of order 6.

THEOREM 2.20 Every group of order less than 6 is abelian.

PROOF: We know that Z_4 and the Klein group are the only groups of order 4, and that each is abelian. The trivial group $\{e\}$ of order 1 is clearly abelian. Any group or order 2 or 3, being of prime order, must be cyclic (Theorem 2.19), and therefore abelian (Theorem 2.6, page 49).

We remind you that $o(a)$ denotes the order of a (Definition 2.5, page 58).

THEOREM 2.21 For any element a in a finite group G:
$$o(a) \big| |G|$$

PROOF: If $o(a) = m$, then $\langle a \rangle = \{a, a^2, \ldots, a^{m-1}, a^m = e\}$ is a subgroup of G consisting of m elements. Consequently: $o(a) \big| |G|$.

THEOREM 2.22 If G is a finite group of order n, then
$$a^n = e \text{ for every } a \in G.$$
(Sum notation: $na = 0$ for every $a \in G$)

PROOF: Let $a \in G$, with $o(a) = m$. Since m divides n (Lagrange's Theorem), $n = tm$ for some $t \in Z^+$. Thus:
$$a^n = a^{tm} = (a^m)^t = e^t = e$$

CHECK YOUR UNDERSTANDING 2.17

Determine the subgroup of the symmetric group S_3:
$$e = \begin{pmatrix} 1 & 2 & 3 \\ 1 & 2 & 3 \end{pmatrix}, \ \alpha_1 = \begin{pmatrix} 1 & 2 & 3 \\ 2 & 3 & 1 \end{pmatrix}, \ \alpha_2 = \begin{pmatrix} 1 & 2 & 3 \\ 3 & 1 & 2 \end{pmatrix}$$
$$\alpha_3 = \begin{pmatrix} 1 & 2 & 3 \\ 1 & 3 & 2 \end{pmatrix}, \ \alpha_4 = \begin{pmatrix} 1 & 2 & 3 \\ 3 & 2 & 1 \end{pmatrix}, \ \alpha_5 = \begin{pmatrix} 1 & 2 & 3 \\ 2 & 1 & 3 \end{pmatrix}$$
generated by the set $\{\alpha_2, \alpha_3\}$.

Answer: S_3

PROOF OF LAGRANGE'S THEOREM

We begin by recalling some material from Chapter 1:

See Definition 1.11 page 29.

An **equivalence relation** $\sim$ on a set X is a relation which is

Reflexive: $x \sim x$ for every $x \in X$,

Symmetric: If $x \sim y$, then $y \sim x$,

Transitive: If $x \sim y$ and $y \sim z$, then $x \sim z$.

See Definition 1.13 page 31.

For $x_0 \in X$ the **equivalence class** of x_0 is the set:
$$[x_0] = \{x \in X | x \sim x_0\}.$$

2.3 Subgroups

LEMMA 2.2 Let H be a subgroup of a group G. The relation $a \sim b$ if $ab^{-1} \in H$ is an equivalence relation on G. Moreover, the equivalence class containing $a \in G$ is the set:
$$[a] = \{ha | h \in H\}$$

PROOF:

$\sim$ is reflexive: $x \sim x$ since $xx^{-1} = e \in H$.

$\sim$ is symmetric:.

$$\boldsymbol{a \sim b} \Rightarrow ab^{-1} = h \text{ for some } h \in H$$
$$\Rightarrow (ab^{-1})^{-1} = h^{-1}$$
Theorem 2.11, page 56: $\Rightarrow (b^{-1})^{-1} a^{-1} = h^{-1}$
Exercise 2, page 60: $\Rightarrow ba^{-1} = h^{-1} \Rightarrow \boldsymbol{b \sim a}$ since $h^{-1} \in H$

(H is a group)

$\sim$ is transitive: If $\boldsymbol{a \sim b}$ and $\boldsymbol{b \sim c}$, then:
$$ab^{-1} \in H \text{ and } bc^{-1} \in H$$
$$\Rightarrow (ab^{-1})(bc^{-1}) \in H$$
$$\Rightarrow a(\boldsymbol{b^{-1} b}) c^{-1} \in H$$
$$\Rightarrow aec^{-1} \in H$$
$$\Rightarrow ac^{-1} \in H \Rightarrow \boldsymbol{a \sim c}$$

Having established the equivalence part of the theorem, we now verify that $[a] = \{ha | h \in H\}$:
$$b \in [a] \Leftrightarrow b \sim a \Leftrightarrow ba^{-1} = h \text{ for some } h \in H$$
$$\Leftrightarrow b = ha \text{ for some } h \in H$$

NOTE: The above set $\{ha | h \in H\}$ will be denoted by Ha, and is said to be a **right coset** of H:
$$Ha = \{ha | h \in H\}$$

We are now in a position to offer a proof of Lagrange's Theorem:

If H is a subgroup of a finite group G, then $|H| \big| |G|$.

PROOF: Theorem 1.13(a), page 32, and Lemma 2.2, tell us that the right cosets of H, $\{Ha | a \in G\}$, partition G. Since G is finite, we can choose $a_1, a_2, ..., a_k$ such that $G = \bigcup_{i=1}^{k} Ha_i$ with $Ha_i \cap Ha_j = \emptyset$ if $i \neq j$.

We now show that each Ha_i has the same number of elements as H, by verifying that the function $f_i: H \to Ha_i$ given by $f_i(h) = ha_i$ is a bijection:

f_i is one-to-one:
$$f_i(h_1) = f_i(h_2) \Rightarrow h_1 a = h_2 a$$
$$\Rightarrow (h_1 a)a^{-1} = (h_2 a)a^{-1}$$
$$\Rightarrow h_1(aa^{-1}) = h_2(aa^{-1}) \Rightarrow h_1 = h_2$$

f_i is onto:
For any given $ha_i \in Ha_i$, $f_i(h) = ha_i$.

Since G is the disjoint union of the k sets $Ha_1, Ha_2, \ldots, Ha_k$, and since each of those sets contains $|H|$ elements: $|G| = k|H|$, and therefore: $|H|\,|\,|G|$.

EXERCISES

Exercise 1-5. Determine if the given subset S is a subgroup of $\langle Z, + \rangle$.

1. $S = \{n \mid n \text{ is even}\}$
2. $S = \{n \mid n \neq 1\}$
3. $S = \{n \mid n \text{ is odd}\}$
4. $S = \{n \mid n \text{ is divisible by 2 and 3}\}$
5. $S = \{n \mid n \text{ is divisible by 2 or 3}\}$

Exercise 6-8. Determine if the given subset S is a subgroup of $\langle Z_8, +_n \rangle$ (see Theorem 2.1, page 42).

6. $S = \{0, 2, 4, 6\}$
7. $S = \{0, 3, 6\}$
8. $S = \{0, 2, 3, 4\}$

Exercise 9-12. Determine if the given subset S is a subgroup of $\langle \Re, + \rangle$.

9. $S = \{x \mid x = 7y \text{ for } y \in \Re\}$
10. $S = \{x \mid x = 7y \text{ for } y \geq 0\}$
11. $S = \{x \mid x = 7 + y \text{ for } y \in \Re\}$
12. $S = \{x \mid x = 7 + y \text{ for } y \geq 0\}$

Exercise 13-18. Determine if the given subset S is a subgroup of $(S_3, \circ)$ where:

$\alpha_0 = \begin{pmatrix} 1 & 2 & 3 \\ 1 & 2 & 3 \end{pmatrix}$, $\alpha_1 = \begin{pmatrix} 1 & 2 & 3 \\ 2 & 3 & 1 \end{pmatrix}$, $\alpha_2 = \begin{pmatrix} 1 & 2 & 3 \\ 3 & 1 & 2 \end{pmatrix}$, $\alpha_3 = \begin{pmatrix} 1 & 2 & 3 \\ 1 & 3 & 2 \end{pmatrix}$, $\alpha_4 = \begin{pmatrix} 1 & 2 & 3 \\ 3 & 2 & 1 \end{pmatrix}$, $\alpha_5 = \begin{pmatrix} 1 & 2 & 3 \\ 2 & 1 & 3 \end{pmatrix}$

13. $S = \{\alpha_0, \alpha_1\}$
14. $S = \{\alpha_0, \alpha_2\}$
15. $S = \{\alpha_0, \alpha_3\}$
16. $S = \{\alpha_0, \alpha_1, \alpha_2\}$
17. $S = \{\alpha_0, \alpha_3, \alpha_5\}$
18. $S = \{\alpha_1, \alpha_2, \alpha_3, \alpha_4, \alpha_5\}$

Exercise 19-21. Determine if the given subset S is a subgroup of $\langle R^3, + \rangle$.

19. $S = \{(a, b, 0) \mid a, b \in \Re\}$
20. $S = \{(a, b, 1) \mid a, b \in \Re\}$
21. $S = \{(a, b, c) \mid c = a + b\}$
22. $S = \{(a, b, c) \mid c = ab\}$

Exercise 23-26. Determine if the given subset S is a subgroup of $\langle M_{2 \times 2}, + \rangle$.

23. $S = \left\{ \begin{bmatrix} a & b \\ a+b & 0 \end{bmatrix} \middle| a, b \in \Re \right\}$
24. $S = \left\{ \begin{bmatrix} a & b \\ a+b & 1 \end{bmatrix} \middle| a, b \in \Re \right\}$
25. $S = \left\{ \begin{bmatrix} a & b \\ a+b & ab \end{bmatrix} \middle| a, b \in \Re \right\}$
26. $S = \left\{ \begin{bmatrix} a & b \\ c & 2a+c \end{bmatrix} \middle| a, b, c \in \Re \right\}$

Exercise 27-30. Determine if the given subset S is a subgroup of $\langle F(\Re), + \rangle$ (see Exercise 49, page 52).

27. $S = \{f \mid f \text{ is continuous}\}$
28. $S = \{f \mid f \text{ is differentiable}\}$
29. $S = \{f \mid f(1) = 1\}$
30. $S = \{f \mid f(1) = 0\}$

Exercise 31-34. Determine if the given subset S is a subgroup of $\langle S_\Re, \circ \rangle$ (see Theorem 2.4, page 46).

31. $S = \{f \mid f \text{ is continuous}\}$
32. $S = \{f \mid f \text{ is differentiable}\}$
33. $S = \{f \mid f(1) = 1\}$
34. $S = \{f \mid f(1) = 0\}$

35. Prove that all subgroups of Z are of the form nZ.

36. Find all subgroups of $\langle Z_6, +_n \rangle$.

37. Prove that if $\{e\}$ and G are the only subgroups of a group G, then G is cyclic of order p, for p prime.

38. Show that a nonempty subset S of a group G is a subgroup of G if and only if
$$s, \bar{s} \in S \Rightarrow s\bar{s}^{-1} \in S$$

39. Show that for any $a, b \in Z^+$, $S = \{na + mb \mid n, m \in Z\}$ is a subgroup of Z.

40. Show that for any group G the set $Z(G) = \{a \in G \mid ag = ga \ \forall g \in G\}$ is a subgroup of G.

41. Let G be an abelian group. Show that for any integer n, $\{a \in G \mid a^n = e\}$ is a subgroup of G.

42. Prove that the subset of elements of finite order in an abelian group G is a subgroup of G (called the **torsion subgroup** of G).

43. Let G be a cyclic group of order n. Show that if m is a positive integer, then G has an element of order m if and only if m divides n.

44. Let a be an element of a group G. The set of all elements of G which commute with a:
$$C(a) = \{b \in G \mid ab = ba\}$$
is called the **centralizer of a in G**. Prove that $C(a)$ is a subgroup of G.

45. Let H be a subgroup of a group G. The **centralizer** $C(H)$ of H is the set of all elements of G that commute with every element of H: $C(H) = \{a \in G \mid ah = ha \text{ for all } h \in H\}$. Prove that $C(H)$ is a subgroup of G.

46. The **center** $Z(G)$ of a group G is the set of all elements in G that commute with ever element of G: $Z(G) = \{a \in G \mid ab = ba \text{ for all } b \in G\}$.

    (a) Prove that $Z(G)$ is a subgroup of G.

    (b) Prove that $a \in Z(G)$ if and only if $C(a) = G$ (see Exercise 43.)

    (c) Prove that $Z(G) = \bigcap_{a \in G} C(a)$.

47. Show that Table C in Figure 2.4, page 45, can be derived from Table B by appropriately relabeling the letters e, a, b, c in B.

48. Let H and K be subgroups of an abelian group G. Verify that $HK = \{hk \mid h \in H \text{ and } k \in K\}$ is a subgroup of G.

49. Let H and K be subgroups of a group G such that $k^{-1}Hk \subseteq H$ for every $k \in K$. Show that $HK = \{hk \mid h \in H \text{ and } k \in K\}$ is a subgroup of G.

50. Prove that H is a subgroup of a group G if and only if $HH^{-1} = \{ab^{-1} \mid a, b \in H\} \subseteq H$.

51. Let H and K be subgroups of an abelian group G of orders n and m respectively. Show that if $H \cap K = \{e\}$, then $HK = \{hk \mid h \in H \text{ and } k \in K\}$ is a subgroup of G of order nm.

52. (a) Prove that the group $\langle Z, +\rangle$ contains an infinite number of subgroups.
    (b) Prove that any infinite group contains an infinite number of subgroups.

53. Let S be a finite subset of a group G. Prove that S is a subgroup of G if and only if $ab \in S$ for every $a, b \in S$.

54. (a) $\{H_i\}_{i=1}^{n}$ be subgroups of a group G. Show that $\bigcap_{i=1}^{n} H_i$ is also a subgroup of G.

    (b) Let $\{H_i\}_{i=1}^{\infty}$ be a collection of subgroups of a group G. Show that $\bigcap_{i=1}^{\infty} H_i$ is also a subgroup of G.

    (c) Let $\{H_\alpha\}_{\alpha \in A}$ be a collection of subgroups of a group G. Show that $\bigcap_{\alpha \in A} H_\alpha$ is also a subgroup of G.

	PROVE OR GIVE A COUNTEREXAMPLE	

55. If H and K are subgroups of a group G, then $H \cup K$ is also a subgroup of G.

56. It is possible for a group G to be the union of two disjoint subgroups of G.

57. In any group G, $\{a \in G | a^n = e \text{ for some } n \in Z\}$ is a subgroup of G.

58. In any abelian group G, $\{a \in G | a^n = e \text{ for some } n \in Z^+\}$ is a subgroup of G.

59. Let G be a group with $a, b \in G$. If $o(a) = n$ and $o(b) = m$, then $(ab)^{nm} = e$.

60. If a group G has only a finite number of subgroups, the G must be finite.

61. If H and K are subgroups of a group G, then $HK = \{hk | h \in H \text{ and } k \in K\}$ is also a subgroup of G.

62. In any group G, $\{a \in G | a^3 = e\}$ is a subgroup of G.

63. No nontrivial group can be expressed as the union of two disjoint subgroups.

§4. HOMOMORPHISMS AND ISOMORPHISMS

Up until now we have focused our attention exclusively on the internal nature of a group G. The time has come to consider links between them:

The word homomorphism comes from the Greek homo meaning "same" and morph meaning "shape."

DEFINITION 2.9
HOMOMORPHISM

A function $\phi: G \to G'$ from a group G to a group G' is said to be a **homomorphism** if $\phi(ab) = \phi(a)\phi(b)$ for every $a, b \in G$.

Let's focus a bit on the equation:

$$\phi(ab) = \phi(a)\phi(b) \quad (*)$$

The operation, ab, on the left side of equation (*) is taking place in the group G while that on the right, $\phi(a)\phi(b)$, occurs in the group G'. What (*) is saying is that you can perform the product in G and then carry the result over to G' (via ϕ), or you can first carry a and b over to G' and then perform the product in that group. Those groups and products, however, need not resemble each other. Consider the following examples:

You can easily verify that $G' = \{-1, 1\}$, under standard multiplication

*	1	−1
1	1	−1
−1	−1	1

is a group.

EXAMPLE 2.5 Let $G = \langle Z, + \rangle$, and let $G' = \{-1, 1\}$ under standard integer multiplication (see margin). Show that $f: G \to G'$ given by:

$$\phi(n) = \begin{cases} 1 & \text{if } n \text{ is even} \\ -1 & \text{if } n \text{ is odd} \end{cases}$$

is a homomorphism.

SOLUTION: We consider three cases:

Case 1. (Both integers are even). If $a = 2n$ and $b = 2m$, then:
$\phi(a + b) = \phi(2n + 2m) = 1$ (since $2n + 2m$ is even)
And also: $\phi(a)\phi(b) = \phi(2n)\phi(2m) = 1 \cdot 1 = 1$.

Case 2. (Both are odd). If $a = 2n + 1$ and $b = 2m + 1$, then:
$\phi(a + b) = \phi[(2n + 1) + (2m + 1)] = \phi(2n + 2m + 2) = 1$
And also:
$\phi(a)\phi(b) = \phi(2n + 1)\phi(2m + 1) = (-1)(-1) = 1$.

Since $\langle Z, + \rangle$ is abelian, we need not consider $a = 2n + 1$ and $b = 2m$,

Case 3. (Even and odd). If $a = 2n$ and $b = 2m + 1$, then:
$\phi(a + b) = \phi[(2n) + (2m + 1)] = \phi[2(n + m) + 1] = -1$
And also: $\phi(a)\phi(b) = \phi(2n)\phi(2m + 1) = (1)(-1) = -1$.

See page 42 for a discussion of the group $\langle Z_n, +_n \rangle$.

EXAMPLE 2.6 Show that the function $\phi: \langle Z, + \rangle \to \langle Z_n, +_n \rangle$ given by $\phi(m) = r$ where $m = nq + r$ with $0 \le r < n$ is a homomorphism.

SOLUTION: Let $a = nq_1 + r_1$, $b = nq_2 + r_2$ with $0 \le r_1 < n$ and $0 \le r_2 < n$, and let $r_1 + r_2 = nq_3 + r_3$ with $0 \le r_3 < n$, then:

$$\begin{aligned}\phi(a+b) &= \phi[(nq_1+r_1)+(nq_2+r_2)]\\ &= \phi[n(q_1+q_2)+(r_1+r_2)]\\ &= \phi[n(q_1+q_2)+(nq_3+r_3)] \text{ with } 0 \le r_3 < n\\ &= \phi[n(q_1+q_2+q_3)+r_3] = r_3 \text{ (since } 0 \le r_3 < n)\end{aligned}$$

And: $\phi(a) +_n \phi(b) = \phi(nq_1+r_1) +_n (\phi nq_2 + r_2)$
$= r_1 +_n r_2 = r_3$ (since $r_1 + r_2 = nq_3 + r_3$ with $0 \le r_3 < n$)

(same)

EXAMPLE 2.7 For any fixed element a in a group G, let $f_a: G \to G$ be given by $f_a(g) = ag$. Show that the function $\phi: G \to \langle S_G, \circ \rangle$ given by $\phi(a) = f_a$ is a one-to-one homomorphism.

See page 46 for a discussion on the symmetric group $\langle S_G, \circ \rangle$.

SOLUTION: ϕ is one-to-one:

$$\phi(a) = \phi(b) \Rightarrow f_a = f_b \underset{\text{in particular}}{\Rightarrow} f_a(e) = f_b(e) \Rightarrow ae = be \Rightarrow a = b$$

To show that ϕ is a homomorphism we need to show that $\phi(ab) = \phi(a) \circ \phi(b)$, which is to say, that the function $f_{ab}: G \to G$ is equal to the function $f_a \circ f_b: G \to G$. Let's do it:

For any $x \in G$: $f_{ab}(x) = (ab)x$

and $(f_a \circ f_b)(x) = f_a[f_b(x)] = f_a(bx) = a(bx)$

By associativity, $(ab)x = a(bx)$, and we are done.

CHECK YOUR UNDERSTANDING 2.18

Show that for any two groups G and G' the function $\phi: G \to G'$ given by $\phi(a) = e$ for every $a \in G$ is a homomorphism (called the **trivial homomorphism** from G to G').

Answer: See page A-11.

Homomorphisms preserve identities, inverses, and subgroups:

THEOREM 2.23 Let $\phi: G \to G'$ be a homomorphism. Then:
(a) $\phi(e) = e'$
(b) $\phi(a^{-1}) = [\phi(a)]^{-1}$
(c) If H is a subgroup of G, then:
$$\phi(H) = \{\phi(h) | h \in H\}$$
is a subgroup of G'.
(d) If H' is a subgroup of G', then:
$$\phi^{-1}(H') = \{h \in H | \phi(h) \in H'\}$$
is a subgroup of G.

PROOF:

(a) Since ϕ is a homomorphism: $\phi(e) = \phi(ee) = \phi(e)(\phi(e))$.

Multiplying both sides by $[\phi(e)]^{-1}$ yields the desired result:

$$[\phi(e)]^{-1}\phi(e) = [\phi(e)]^{-1}[\phi(e)\phi(e)]$$
$$e' = ([\phi(e)]^{-1}[\phi(e)])\phi(e)$$
$$e' = e'\phi(e) = \phi(e)$$

(b) Since $\phi(a^{-1})\phi(a) = \phi(a^{-1}a) = \phi(e) \underset{(a)}{=} e'$:

$$\phi(a^{-1}) = [\phi(a)]^{-1}.$$

(c) We use Theorem 2.14, page 62, to show that the nonempty set $\phi(H)$ is a subspace of G':

Since $\phi(a)\phi(b) = \phi(ab)$: $\phi(H)$ is closed with respect to the operation in G'.

Since, for any $a \in G$, $\phi(a^{-1}) = [\phi(a)]^{-1}$:

$$[\phi(a)]^{-1} \in \phi(H) \text{ for every } \phi(a) \in \phi(H).$$

(d) We use Theorem 2.14 to show that the nonempty set $\phi^{-1}(H')$ is a subspace of G:

Let $a, b \in \phi^{-1}(H')$. To say that $ab \in \phi^{-1}(H')$ is to say that $\phi(ab) \in H'$, and it is:

Since $\phi(ab) = \phi(a)\phi(b)$, and since H', being a subgroup of G', is closed with respect to the operations in G': $\phi(ab) \in H'$.

Let $a \in \phi^{-1}(H')$. To say that $a^{-1} \in \phi^{-1}(H')$ is to say that $\phi(a^{-1}) \in H'$, and it is:

Since $\phi(a^{-1}) = [\phi a]^{-1}$, and since H' contains the inverse of each of its elements: $\phi(a^{-1}) \in H'$.

CHECK YOUR UNDERSTANDING 2.19

Let $\phi: G \to G'$ and $\theta: G' \to G''$ be homomorphisms. Prove that the composite function $\theta \circ \phi: G \to G''$ is also a homomorphism.

Answer: See page A-11.

2.4 Homomorphisms and Isomorphisms

IMAGE AND KERNEL

For any given homomorphism $\phi: G \to G'$, we define the kernel of ϕ to be the set of elements in G which map to the identity $e' \in G'$ [see Figure 2.7(a)]. We define the set of all elements in G' which are "hit" by some $\phi(a)$ to be the image of ϕ [see Figure 2.7(b)].

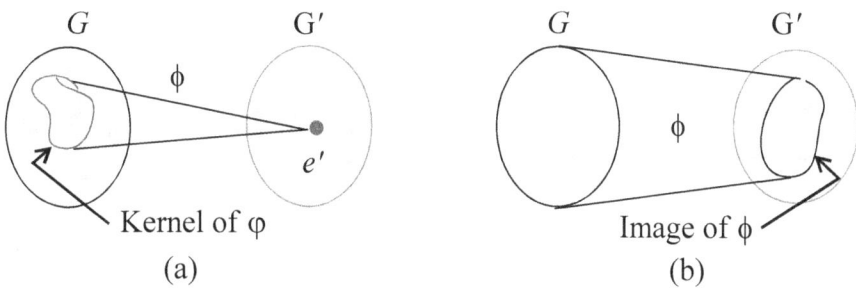

Kernel of φ   Image of φ
(a)   (b)

Figure 2.7

More formally:

Utilizing the notation of Definition 1.3, page 2:
$Ker(\phi) = \phi^{-1}[\{e'\}]$
$Im(\phi) = \phi[G]$

DEFINITION 2.10   Let $\phi: G \to G'$ be a homomorphism.

KERNEL   The **kernel** of ϕ, denoted by $Ker(\phi)$, is given by:
$$Ker(\phi) = \{a \in G | \phi(a) = e'\}$$

IMAGE   The **image** of ϕ, denoted by $Im(\phi)$, is given by:
$$Im(\phi) = \{\phi(a) | a \in G\}$$

Both the kernel and image of a homomorphism turn out to be subgroups of their respective groups:

THEOREM 2.24   Let $\phi: G \to G'$ be a homomorphism. Then:

(a) $Ker(\phi)$ is a subgroup of G.

(b) $Im(\phi)$ is a subgroup of G'.

PROOF: (a) A consequence of Theorem 2.22(d) and the fact that $\{e'\}$ is a subgroup of G'

(b) A consequence of Theorem 2.2(c).

CHECK YOUR UNDERSTANDING 2.20

Show that the function $\phi: 2\mathbb{Z} \to 4\mathbb{Z}$ given by $\phi(2n) = 8n$ is a homomorphism. Determine the kernel and image of ϕ.

Answer: See page A-11.

Definition 2.10 tells us that a homomorphism $\phi: G \to G'$ is onto if and only if $\text{Im}(\phi) = G'$. The following result is a bit more interesting, in that it asserts that in order for a homomorphism to be one-to-one, it need only behave "one-to-one-ish" at e (see margin):

THEOREM 2.25 A homomorphism $\phi: G \to G'$ is one-to-one if and only if $\text{Ker}(\phi) = \{e\}$.

A homomorphism $\phi: G \to G'$ must map e to e'. What this theorem is saying is that if e is the only element that goes to e', then no element of G' is going to be hit by more that one element of G. This is certainly not true for arbitrary functions:

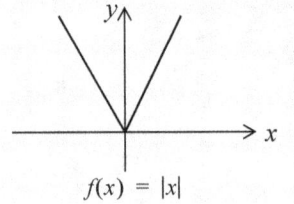

$f(x) = |x|$

PROOF: Suppose ϕ is one-to-one. If $a \in \text{Ker}(\phi)$, then both $\phi(a) = e'$ and $\phi(e) = e'$ [Theorem 2.22(a)]. Consequently $a = e$ (since ϕ is assumed to be one-to-one). Hence: $\text{Ker}(\phi) = \{e\}$.

Conversely, assume that $\text{Ker}(\phi) = \{e\}$. We need to show that if $\phi(a) = \phi(b)$, then $a = b$. Let's do it:

$$\phi(a) = \phi(b)$$
$$\phi(a)[\phi(b)]^{-1} = e'$$

Theorem 2.22(b): $\quad \phi(a)\phi(b^{-1}) = e'$

ϕ is a homomorphism: $\quad \phi(ab^{-1}) = e'$

$\text{Ker}(\phi) = \{e\}$: $\quad ab^{-1} = e$

$$(ab^{-1})b = eb$$
$$a = b$$

CHECK YOUR UNDERSTANDING 2.21

Let $\phi: G \to G'$ be a homomorphism. Show that if there exists an element $c \in G$ (not necessarily the identity e) such that if $\phi(c) = \phi(a)$ then $c = a$, then ϕ is one-to-one.

In other words: for a homomorphism $\phi: G \to G'$ to be one-to-one, it need only behave "one-to-one-ish" at any one-point in G."

Answer: See page A-11.

The word *isomorphism* comes from the Greek *iso* meaning "equal" and *morph* meaning "shape."

ISOMORPHISMS

As previously noted, a homomorphism $\phi: G \to G'$ preserves the algebraic structure in that $\phi(ab) = \phi(a)\phi(b)$. An isomorphism also preserves set structures, in that it pairs of the elements of the set G with those of the set G'. More formally:

DEFINITION 2.11

ISOMORPHISM A homomorphism $\phi: G \to G'$ which is also a bijection is said to be an **isomorphism** from the group G to the group G'.

AUTOMORPHISM An isomorphism $\phi: G \to G$ is said to be an **automorphism on G**.

ISOMORPHIC Two groups G and G' are **isomorphic**, written $G \cong G'$, if there exists an isomorphism from one of the groups to the other.

2.4 Homomorphisms and Isomorphisms

EXAMPLE 2.8 Show that the group $\langle \Re, + \rangle$ of real numbers under addition is isomorphic to the group $\langle \Re^+, \cdot \rangle$ of **positive** real numbers under multiplication.

In this discussion we are not using e to denote the identity element in $\langle \Re^+, \cdot \rangle$ (which is 1). Here, e is the transcendental number $e \approx 2.718$.

SOLUTION: We show that the function $\phi: \langle \Re, + \rangle \to \langle \Re^+, \cdot \rangle$ given by $\phi(a) = e^a$ is an isomorphism:

Homomorphism:
$$\phi(a+b) = e^{a+b} = e^a e^b = \phi(a)\phi(b)$$

One-to-one: (See Theorem 2.24)

The identity in $\langle \Re^+, \cdot \rangle$ ⌐ ⌐ The identity in $\langle \Re, + \rangle$
$$\phi(a) = 1 \Rightarrow e^a = 1 \Rightarrow a = 0$$

Onto: For $a \in \langle \Re^+, \cdot \rangle$, we have: $\phi(\ln a) = e^{\ln a} = a$.

CHECK YOUR UNDERSTANDING 2.22

(a) Prove that $\cong$ is an equivalence relation on any set of groups (see Definition 1.12, page 29).
(b) Prove that $nZ \cong mZ$ for any $n, m \in Z^+$.
(c) Let $g \in G$. Prove that the map $i_g: G \to G$ given by
$$i_g(x) = gxg^{-1} \ \forall x \in G$$
is an automorphism (called an **inner automorphism**.)

Answer: See page A-12.

Algebraically speaking, there is but one cyclic group of order n, and but one infinite cyclic group:

THEOREM 2.26 (a) If the cyclic group $G = \langle a \rangle$ is of order n, the $G \cong \langle Z_n, +_n \rangle$.

(b) If $G = \langle a \rangle$ is infinite, the $G \cong \langle Z, + \rangle$.

PROOF: (a) We show that the function:
$$\phi: \{0, 1, 2, \ldots, n-1\} \to \{\underset{e}{a^0}, a^1, a^2, \ldots, a^{n-1}\}$$

given by $\phi(i) = a^i$ is an isomorphism from $\langle Z_n, +_n \rangle$ to $G = \langle a \rangle$:

One-to-one. For $0 \le i \le j < n$:
$$\phi(i) = \phi(j) \Rightarrow a^i = a^j \Rightarrow a^{i-j} = a^0 \Rightarrow i-j = 0 \Rightarrow i = j$$

Onto. For $a^i \in \{a^0, a^1, a^2, \ldots, a^{n-1}\}$, $\phi(i) = a^i$

Homomorphism: $\phi(i+j) = a^{i+j} = a^i a^j = \phi(i)\phi(j)$

(b) Your turn:

CHECK YOUR UNDERSTANDING 2.23

Show that every infinite cyclic group is isomorphic to $\langle Z, + \rangle$.

Answer: See page A-12.

A ROSE BY ANY OTHER NAME

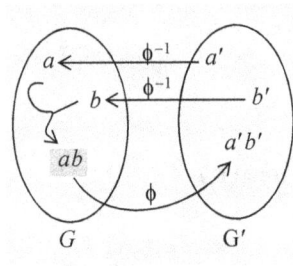

Let $\phi: G \to G'$ be an isomorphism. Being a bijection it links every element in G with a unique element in G' (every element in G has its own G' counterpart, and vice versa). Moreover, if you know how to function algebraically in G, then you can also figure out how to function algebraically in G' (and vice versa). Suppose, for example, that you forgot how to multiply in the group G', but remember how to multiply in G. To figure out $a'b'$ in G' you can take the "ϕ^{-1}-bridge" back to G to find **the** elements a and b for which $\phi(a) = a'$ and $\phi(b) = b'$, perform the product ab in G, and then take the "ϕ-bridge" back to G' to determine the product $a'b'$: $\phi(ab)$.

Basically, if a group G is isomorphic to G', then the two groups can only differ in appearance, but not algebraically. Consider, for example, the two groups which previously appeared in Figure 2.1, page 43:

Z_4:

$+_4$	0	1	2	3
0	0	1	2	3
1	1	2	3	0
2	2	3	0	1
3	3	0	1	2

(a)

K:

*	e	a	b	c
e	e	a	b	c
a	a	e	c	b
b	b	c	e	a
c	c	b	a	e

(b)

Both contain four elements ($\{0,1,2,3\}$ and $\{e,a,b,c\}$); so, as far as sets are concerned, they "are one and the same" (different element-names, that's all). But as far as groups go, they are **not** the same (not isomorphic). Here are two algebraic differences (either one of which would serve to prove that the two groups are not isomorphic):

1. Z_4 is cyclic while the Klein 4-group, K, is not.
2. There exist three elements in K of order 2 (see Definition 2.5, page 58), while Z_4 contains but one (the element 2).

To better substantiate the above claims:

THEOREM 2.27 If $G \cong G'$, then:

(a) G is cyclic if and only if G' is cyclic.

(b) For any given integer n, there exists an element $a \in G$ such that $a^n = e$ if and only if there exists an element $a' \in G'$ such that $(a')^n = e'$.

PROOF: Let $\phi\colon G \to G'$ be an isomorphism.

(a) Suppose G is cyclic, with $G = \langle a \rangle$. We show $G' = \langle \phi(a) \rangle$ by showing that for any $b' \in G'$, there exists $n \in Z$ such that $b' = [\phi(a)]^n$:

Let $b \in G$ be such that $\phi(b) = b'$. Since $G = \langle a \rangle$, there exists $n \in Z$ such that $b = a^n$. Then:
$$b' = \phi(b) = \phi[a^n] \underset{\substack{\uparrow \\ \text{Exercise 16}}}{=} [\phi(a)]^n$$

The "only-if" part follows from the fact that if G is isomorphic to G', then G' is isomorphic to G [see CYU 2.23(a)].

(b) Let $a \in G$ be such that $a^n = e$. Then:
$$[\phi(a)]^n = \phi(a^n) = \phi(e) = e'$$
The "only-if" part follows from CYU 2.22(a).

CHECK YOUR UNDERSTANDING 2.24

Answer: See page A-12.

Prove that if $G \cong G'$, then G is abelian if and only if G' is abelian.

A property of a group G that is shared by all groups isomorphic to G is said to be a **group invariant** property. For example, abelian and cyclic are group invariant properties. Other group invariant properties are cited in the exercises.

In general, one can show that two groups are not isomorphic by exhibiting a group invariant property that holds in one of the groups but not in the other. For example, the permutation group S_3 is not isomorphic to $\langle Z_6, +_6 \rangle$ as one is abelian while the other is not.

The following results underlines the importance of symmetric groups (see discussion on page 46).

Arthur Cayley (1821-1885)

THEOREM 2.28 (Cayley)   Every group is isomorphic to a subgroup of a symmetric group.

PROOF: The function $\phi\colon G \to S_G$ given by
$$\phi(g) = f_g\colon G \to G \text{ where } f_g(x) = gx \ (\forall x \in G)$$
was shown to be a one-too-one homomorphism in Example 2.7.

Since ϕ it is onto the subspace $\phi(G)$ of S_G:
$$\phi\colon G \to \phi(G) \text{ is an isomorphism.}$$

EXERCISES

Exercise 1-13. Show that the given function $\phi: G \to G'$ from the group G to the group G' is a homomorphism.

1. $G = G' = \langle Z, + \rangle$ and $\phi(n) = 2n$.
2. $G = G' = \langle Z, + \rangle$ and $\phi_r(n) = rn$ for $r \in \Re$.
3. $G = \langle Z, + \rangle$, $G' = \langle \Re, + \rangle$ and $\phi(n) = n$.
4. $G = \langle Z, + \rangle$, $G' = Z_3$ and $\phi(n) = r$ where $n = 3m + r$ with $0 \le r < 3$.
5. $G = \langle Z, + \rangle$, $G' = \langle \{-1, 1\}, \cdot \rangle$ and $\phi(n) = 1$ if n is even and $f(n) = -1$ if n is odd.
6. $G = Z_5$, $G' = Z_2$ and $\phi(n) = r$ where $n = 2d + r$ with $0 \le r < 2$.
7. $G = Z_6$, $G' = Z_2$ and $\phi(n) = r$ where $n = 2d + r$ with $0 \le r < 2$.
8. $G = G' = \langle M_{2 \times 2}, + \rangle$ and $\phi\left(\begin{bmatrix} a & b \\ c & d \end{bmatrix}\right) = \begin{bmatrix} a+b & d \\ -c & 0 \end{bmatrix}$.
9. $G = \langle M_{2 \times 2}, + \rangle$, $G' = \Re$ and $\phi\left(\begin{bmatrix} a & b \\ c & d \end{bmatrix}\right) = ad - bc$.
10. $G = S_n$, $G' = Z_2$ and $\phi(\sigma) = \begin{cases} 0 & \text{if } \sigma \text{ is an even permutation} \\ 1 & \text{if } \sigma \text{ is an even permutation} \end{cases}$.
11. $G = S_3$, $G' = S_4$ and $[\phi(\sigma)](i) = \begin{cases} \sigma(i) & \text{if } i < 4 \\ 4 & \text{if } i = 4 \end{cases}$,
12. $G = G'$ with G abelian, and $\phi(a) = a^{-1}$ for $a \in G$.
13. $G = G'$ with G abelian, $n \in Z^+$, and $\phi(a) = a^n$ for $a \in G$.

Exercise 14-16. Show that the given function $\phi: G \to G'$ from the group G to the group G' is not a homomorphism.

14. $G = G' = \langle Z, + \rangle$ and $\phi(n) = n + 1$.
15. $G = G' = \langle M_{2 \times 2}, + \rangle$ and $\phi\left(\begin{bmatrix} a & b \\ c & d \end{bmatrix}\right) = \begin{bmatrix} a+b & d \\ -c & 1 \end{bmatrix}$.
16. $G = G' = S_3$ and $\phi(a) = a^{-1}$ for $a \in G$.

Exercise 14-16. Find the kernel of the homomorphism of:

17. Exercise 1.   18. Exercise 2.   19. Exercise 3.
20. Exercise 4.   21. Exercise 5.   22. Exercise 6.
23. Exercise 7.   24. Exercise 8.   25. Exercise 9.
26. Exercise 10. 27. Exercise 11. 28. Exercise 12.

29. Let $\langle \Re, + \rangle$ denote the group of all real numbers under addition, and $\langle \Re^+, \cdot \rangle$ the group of all positive real numbers under multiplication. Show that the map $\phi: \Re^+ \to \Re$ given by $\phi(x) = \ln x$ is an isomorphism.

30. Let $\phi: G \to G'$ be a homomorphism and let $a \in G$. Prove that $\phi(a^n) = [\phi(a)]^n$ for every $n \in Z$.

31. Let $\phi: G \to G'$ be a homomorphism and let $a \in G$. Show that the map $\phi: Z \to G$ given by $\phi(n) = a^n$ is a homomorphism.

32. Let $\phi: G \to G'$ be a homomorphism with G' finite. Show that $|\phi(G)|$ is a divisor of $|G'|$.

33. Let $\phi: G \to G'$ be a homomorphism. Prove that for all $a, b \in G$:
$$\phi(ab^{-1}) = \phi(a)\phi(b)^{-1} \text{ and } \phi(a^{-1}b) = \phi(a)^{-1}\phi(b)$$

34. Let $\phi: G \to G'$ be a homomorphism, Show that:
    (a) If ϕ is onto and if G is abelian, then G' is abelian.

    (a) If ϕ is one-to-one and if G' is abelian, then G is abelian.

35. Prove that a group G is abelian if and only if the function $f: G \to G$ given by $f(g) = g^{-1}$ is a homomorphism.

36. Let $G = \langle a \rangle$ be cyclic and let G' be any group. Let $\phi: G \to G'$ be a homomorphism. Prove that $\text{Im}(\phi)$ is cyclic.

37. Let $\phi: G \to G'$ be a homomorphism. Show that if $k \in \text{Ker}(\phi)$, then $gkg^{-1} \in \text{Ker}(\phi)$ for every $g \in G$.

38. Let G, G', and G'' be groups. Show that if $\phi: G \to G'$ and $\gamma: G' \to G''$ are homomorphisms, then so is $\gamma \circ \phi: G \to G''$.

39. Let $\phi: G \to G'$ be a homomorphism. Show that $\phi(G)$ is abelian if and only if for all $a, b \in G: aba^{-1}b^{-1} \in \text{Ker}(\phi)$

40. Let $\phi: G \to G'$ be a homomorphism. Prove that, for any given $x \in G$:
$$\{g \in G | \phi(g) = \phi(x)\} = \{xk | k \in \text{Ker}(\phi)\}$$

41. Let $G = \langle a \rangle$ be cyclic and let H be any group. Prove that for any chosen $h \in H$ there exists a unique homomorphism $\phi: G \to H$ such that $\phi(a) = h$.
    So, a homomorphism on a cyclic group $G = \langle a \rangle$ is completely determined by its action on a.

42. Let $\phi: G \to G'$ be a homomorphism. Prove that, for any given $x \in G$:
$$\{g \in G | \phi(g) = \phi(x)\} = \{xk | k \in \text{Ker}(\phi)\}$$

43. Let A, B, C, and D be groups. Show that if $A \cong B$ and $C \cong D$, then $A \times C \cong B \times D$ (see Exercise 52, page 52).

44. Let G and G' be groups. Show that $G \times G' \cong G' \times G$ (see Exercise 52, page 52).

45. (a) Show that the set $Z \times Z = \{(a, b) | a, b \in Z\}$, with $(a, b) * (c, d) = (a + c, b + d)$ is a group.

   (b) Verify that the functions $\phi_1: Z \times Z \to Z$ and $\phi_2: Z \times Z \to Z$ given by $\phi_1(a, b) = a$ and $\phi_2(a, b) = b$, respectively, are homomorphisms.

   (c) Show that the function $\phi: Z \times Z \to Z$ given by $\phi(a, b) = 2\phi_1(a, b) + 3\phi_2(a, b)$ is a homomorphism.

   (d) Show that the function $\theta: Z \times Z \to Z \times Z$ given by $\theta(a, b) = [\phi_2(a, b), \phi_1(a, b)]$ is an isomorphism.

46. For $m \in Z$, $m \neq 0$, let $\phi_m: Z \to Z$ be given by $\phi_m(n) = mn$.

   (a) Show that ϕ_m is a one-to-one homomorphism.

   (b) Show that ϕ_m is an isomorphism if and only if $m = \pm 1$.

47. Let $F(\Re)$ denote the additive group of real valued function (see Exercise 49, page 52), and let $\Re$ denote the additive group of real numbers. Prove that for any $c \in \Re$ the function $\phi_c: F(\Re) \to \Re$ given by $\phi_c(f) = f(c)$ for $f \in F(\Re)$ is a homomorphism (called an **evaluation homomorphism**.)

48. Let $D(\Re)$ denote the set of differentiable functions from $\Re$ to $\Re$.

   (a) Show that $\langle D(\Re), + \rangle$ is a group.

   (b) Show that for any $c \in \Re$ the function $\phi_c: D(\Re) \to \Re$ given by $\phi_c(f) = f(c)$ is a homomorphism.

   (c) Is ϕ_c one-to-one for any c?

   (d) Is ϕ_c onto for any c?

49. Let $C(\Re)$ denote the set of continuous real valued functions.

   (a) Show that $\langle C(\Re), + \rangle$ is a group.

   (b) Show that for any closed interval $[a, b]$ in $\Re$ the function $\phi: C(\Re) \to \Re$ given by $\phi(f) = \int_a^b f(x) dx$ is a homomorphism.

   (c) Show that the function $\theta: C(\Re) \to \Re$ given by $\phi(f) = \int_0^1 f(x) dx + 2 \int_2^3 f(x) dx$ is a homomorphism.

50. Show that for any $\tau \in S_3$, the function $\phi: S_3 \to S_3$ given by $\phi(\sigma) = \sigma \circ \tau$ is a homomorphism. Is it necessarily an isomorphism?

51. Let G be a group. Prove that $\text{Aut}(G) = \langle \{\phi | \phi: G \to G \text{ is an automorphism}\}, \circ \rangle$ is a group.

Exercise 34-40. Show that the give property on a G is an invariant.

52. $|G|$ — the order of a finite group G.

53. G contains a nontrivial cyclic subgroup.

54. G contains an element of order n for given $n \geq 1$.

55. G contains m elements of order n for given $n \geq 1$.

56. G contains a subgroup of order of order n for given $n \geq 1$.

57. The number of elements in $T_n = \{g \in G | o(g) = n\}$ (see Definition 2.5, page 58).

58. The number of elements in $Z(G)$ — the center of a finite group G. (See Exercise 45, page 70.)

	PROVE OR GIVE A COUNTEREXAMPLE	

59. The additive group $\Re$ is isomorphic to the additive group Q of rational numbers)

60. The additive group Z is isomorphic to the additive group Q of rational numbers)

61. If ϕ is a homomorphism from a group G to a cyclic group $G' = \langle a \rangle$, then $\text{Ker}(\phi)$ is a cyclic subgroup of G.

62. If ϕ is an isomorphism from a group G to a cyclic group $G' = \langle a \rangle$, then $\text{Ker}(\phi)$ is a cyclic subgroup of G.

63. For $C(\Re)$ the group of continuous real valued functions under addition the function $\phi: C(\Re) \to \Re$ given by $\phi(f) = \left(\int_0^1 f(x)dx\right)\left(\int_2^3 f(x)dx\right)$ is a homomorphism.

64. If $n \neq m$, S_n and S_m are not isomorphic.

§5. SYMMETRIC GROUPS

Cayle's Theorem asserts that every group is isomorphic to a subgroup of a symmetric group. It follows that if one knew everything about symmetric groups, then one would know everything about groups in general. Alas, however, symmetric groups $\langle S_X, \circ \rangle$ are not "easy to own," especially if X is an infinite set.

In this section we focus our attention on finite symmetric groups, specifically on the groups S_n of section 2.1 (see page 46).

CYCLE DECOMPOSITION

Consider the permutation:

$$\sigma = \begin{pmatrix} 1 & 2 & 3 & 4 & 5 & 6 & 7 \\ 3 & 5 & 4 & 7 & 2 & 6 & 1 \end{pmatrix}$$

To get a sense of its action, let's use the symbol $1 \to 3$ to indicate that σ maps 1 to 3. We then have:

$1 \to 3 \to 4 \to 7 \to 1$; or, better yet:

Adhering to convention we let the symbol (1,3,4,7) represent the permutation in S_7 that acts like σ on the integers 1, 3, 4, and 7, and leaves 2, 5, and 6 fixed:

$$(1, 3, 4, 7) = \begin{pmatrix} 1 & 2 & 3 & 4 & 5 & 6 & 7 \\ 3 & 2 & 4 & 7 & 5 & 6 & 1 \end{pmatrix}$$

In general, a cycle of the form $(n_1, n_2, ..., n_k)$ is said to be a *k*-cycle.

(said to be a **4-cycle** of the permutation σ)

Proceeding as above, but starting with 2 (or 5) we arrive at the 2-cycle:

$$(2, 5) = \begin{pmatrix} 1 & 2 & 3 & 4 & 5 & 6 & 7 \\ 1 & 5 & 3 & 4 & 2 & 6 & 7 \end{pmatrix}$$

All that remains is 6. But 6 is stationary under σ, so;

$$(6) = \begin{pmatrix} 1 & 2 & 3 & 4 & 5 & 6 & 7 \\ 1 & 2 & 3 & 4 & 5 & 6 & 7 \end{pmatrix} \leftarrow \text{the identity permutation}$$

In writing a permutation $\sigma \in S_n$ as a product of cycles, we generally don't include cycles of length 1, as any such cycle is the identity in S_n. In particular, it is understood that the permutation
$\sigma = (1, 3, 4, 7)(2, 5) \in S_7$
leaves 6 fixed.

At this point we can express σ as a product of cycles; specifically:

$$\sigma = (1, 3, 4, 7)(2, 5) \text{ i.e:}$$

$$\begin{pmatrix} 1 & 2 & 3 & 4 & 5 & 6 & 7 \\ 3 & 5 & 4 & 7 & 2 & 6 & 1 \end{pmatrix} = \begin{pmatrix} 1 & 2 & 3 & 4 & 5 & 6 & 7 \\ 3 & 2 & 4 & 7 & 5 & 6 & 1 \end{pmatrix} \begin{pmatrix} 1 & 2 & 3 & 4 & 5 & 6 & 7 \\ 1 & 5 & 3 & 4 & 2 & 6 & 7 \end{pmatrix}$$

Since (1, 3, 4, 7) does not move 2 or 5, and since (2,5) does not effect 1, 3, 4, or 7, the two cycles are said to be **disjoint** and must commute (see Exercise 21):

$$(1, 3, 4, 7)(2, 5) = (2, 5)(1, 3, 4, 7)$$

2.5 Symmetric Groups

Just as any integer can be expressed as a product of primes, so then can permutations be expressed as products of cycles.

In general:

THEOREM 2.29 Every permutation in S_n can be expressed as a product of disjoint cycles.

PROOF: (By induction on n)

Let $P(n)$ be the proposition that every permutation in S_n can be expressed as a product of disjoint cycles.

I. $P(1)$ is true: $S_1 = \binom{1}{1} = (1)$.

II. Assume $P(m)$ is true for $1 \leq m \leq k$ (see API, page 16)

III. We show that $P(k+1)$ is true, thereby completing the proof:
Let $\sigma \in S_{k+1}$. Since S_{k+1} is a finite group, σ has finite order, say $o(\sigma) = m$.

Note that $\sigma^m(1)$ cycles back to 1.

If $m = k+1$, then $\sigma = (1, \sigma(1), \sigma^2(1), ..., \sigma^{m-1}(1))$ — a cycle.

If $m < k+1$, then consider the set

$$O_\sigma(1) = \{1, \sigma(1), \sigma^2(1), ..., \sigma^{m-1}(1)\}$$

(called the **orbit** of 1 under σ)

Pulling the above orbit out from $\{1, 2, ..., k+1\}$:

$$\{1, 2, ..., k+1\} - O_\sigma(1)$$

we arrive at a permutation σ_s on a set of $s = (k+1) - m$ elements, with $1 \leq s \leq k$. By II, σ_s can be written as a product of disjoint cycles $\tau_1, \tau_2, ..., \tau_t$. It follows that:

$$\sigma = \tau_1 \cdot \tau_2 \cdots \tau_t \cdot O_\sigma(1)$$

(note that the orbit $O_\sigma(1)$ is disjoint from all of the τ_i's)

In the next example we again focus on the cycle-decomposition-procedure, but in reverse.

EXAMPLE 2.9 Construct a permutation $\sigma \in S_{10}$ that can be expressed as a product of a 2-cycle, a 3-cycle, and a 4-cycle.

SOLUTION: Any such permutation must leave $10 - (2 + 3 + 4) = 1$ element fixed. We decide to go with the element 3 [see Figure 2.8(a)].

$$\begin{pmatrix} 1 & 2 & 3 & 4 & 5 & 6 & 7 & 8 & 9 & 10 \\ & & 3 & & & & & & & \end{pmatrix}$$
(a)

$$\begin{pmatrix} 1 & 2 & 3 & 4 & 5 & 6 & 7 & 8 & 9 & 10 \\ & & 3 & & & & & & 9 & 7 \end{pmatrix}$$
(b)

$$\begin{pmatrix} 1 & 2 & 3 & 4 & 5 & 6 & 7 & 8 & 9 & 10 \\ 2 & 4 & 3 & 1 & & & & & 9 & 7 \end{pmatrix}$$
(c)

$$\begin{pmatrix} 1 & 2 & 3 & 4 & 5 & 6 & 7 & 8 & 9 & 10 \\ 2 & 4 & 3 & 1 & 10 & 5 & 9 & 6 & 7 & 8 \end{pmatrix}$$
(d)

Figure 2.8

We then choose 7, along with 9, to generate the 2-cycle (7, 9) [see Figure 2.8(b)]. Of the remaining 7 elements we decide to go with 1, 2, and 4 to create the 3-cycle (1, 2, 4) [Figure 2.8(c)]. All that's left are the elements 5, 6, 8, 10, and decide to mold them into the cycle (8, 6, 5, 10) — bringing us to the completed permutation $\sigma \in S_{10}$ in Figure 2.8(d) with cycle decomposition:

$$\sigma = (7, 9)(1, 2, 4)(8, 6, 5, 10)$$

CHECK YOUR UNDERSTANDING 2.25

(a) Express $\sigma = \begin{pmatrix} 1 & 2 & 3 & 4 & 5 & 6 & 7 & 8 & 9 & 10 \\ 3 & 9 & 4 & 1 & 5 & 6 & 2 & 7 & 8 & 10 \end{pmatrix}$ as a product of cycles.

(b) Construct a permutation $\sigma \in S_{10}$ that can be expressed as a product of two 2-cycles and a 5-cycle.

Answer:
(a) $(1, 3, 4)(2, 9, 8, 7)$
(b) See page A-13.

ORDER OF PERMUTATIONS

We already know that the symmetric group S_n has order $n!$. We now turn our attention to the task of determining the order of permutations in S_n. Let's start off by considering the 4-cycle:

$$\sigma = (2, 6, 3, 5) \in S_n \text{ for } n \geq 6$$

Focusing on the element 2 we have:

$$\sigma(2) = 6, \ \sigma^2(2) = 3, \ \sigma^3(2) = 5, \ \sigma^4(2) = 2$$

So, the smallest power s of σ such that $\sigma^s(2) = 2$ is $s = 4$, and the same can be said for the elements 6, 3, and 5. It follows, since all of the remaining elements in $\{1, 2, ..., n\}$ are held fixed by σ, that $o(\sigma) = 4$. Indeed, as you are invited to establish in the exercises:

THEOREM 2.30 *Every k-cycle in S_n has order k.*

Moving things along we reconsider the permutation

$$\sigma = \begin{pmatrix} 1 & 2 & 3 & 4 & 5 & 6 & 7 & 8 & 9 & 10 \\ 2 & 4 & 3 & 1 & 10 & 5 & 9 & 6 & 7 & 8 \end{pmatrix} = (7, 9)(1, 2, 4)(8, 6, 5, 10)$$

that surfaced in Example 2.9. We know, from Theorem 2.30, that:

$$o[(7, 9)] = 2, \ o[(1, 2, 4)] = 3, \text{ and } o[(8, 6, 5, 10)] = 4$$

It follows that $\sigma^s = e$ for any s that is divisible by 2, 3, and 4. In particular, $s = \text{lcm}(2, 3, 4) = 12$ will work. Moreover, since the three cycles are disjoint, no positive integer smaller that $\text{lcm}(2, 3, 4)$ will do the trick, bringing us to:

2.5 Symmetric Groups 87

THEOREM 2.31 If $\sigma \in S_n$ has a cycle decomposition of disjoint cycles of order (length) $k_1, k_2, \ldots, k_s$, then:

$$o(\sigma) = \text{lcm}(k_1, k_2, \ldots, k_s)$$

Note the "**disjoint** cycles" condition in the theorem.
A case in point:
The 2-cycles $(1, 2), (1, 3)$ are not disjoint, and their product is not of order 2:
$o[(1, 2)(1, 3)]$
$= o(1, 3, 2) = 3$

Answer: 6

CHECK YOUR UNDERSTANDING 2.26

Determine the order of the permutation:

$$\sigma = \begin{pmatrix} 1 & 2 & 3 & 4 & 5 & 6 & 7 & 8 \\ 3 & 8 & 6 & 7 & 4 & 1 & 5 & 2 \end{pmatrix}$$

A cycle in S_n is, in a sense, a primitive object in that it cannot be decomposed into a product of smaller **disjoint** cycles. It can, however, always be decomposed into a product of 2-cycles, called **transpositions**. Consider, for example the cycle $\sigma = (3, 2, 5, 1)$:

While the above 4-cycle could reside in any S_n with $n \geq 5$, all elements other that 1, 2, 3, and 5 are immune to its action. That being the case, we might as well embed it in S_5. We then have:

Note that the decomposition of a cycle as a product of transposition is **not unique**.
A case in point:
$(3, 2, 5, 1) = (3, 1)(3, 5)(3, 2)$
$= (3, 1)(3, 5)(3, 2)(1, 2)(1, 2)$

$$\overline{(3,2)} \downarrow \overline{(3,5)} \downarrow \overline{(3,1)} \downarrow \begin{pmatrix} 1\ 2\ 3\ 4\ 5 \\ 1\ 3\ 2\ 4\ 5 \\ 1\ 5\ 2\ 4\ 3 \\ 3\ 5\ 2\ 4\ 1 \end{pmatrix} \Rightarrow \sigma = (3,1)(3,5)(3,2)$$

Note the pattern:
first switch 3 with 2
then switch 3 with 5
finally switch 3 with 1

Generalizing the above pattern, we have (Exercise 27):

THEOREM 2.32 Any cycle can be expressed as a product of transpositions.

Merging the above result with Theorem 2.29 we come to

THEOREM 2.33 Every permutation can be expressed as a product of transpositions.

CHECK YOUR UNDERSTANDING 2.27

(a) Express th cycle $(3, 1, 6, 4, 7)$ as a product of transpositions.

(b) Express the permutation $\begin{pmatrix} 1 & 2 & 3 & 4 & 5 & 6 & 7 & 8 & 9 & 10 \\ 2 & 4 & 3 & 1 & 10 & 5 & 9 & 6 & 7 & 8 \end{pmatrix}$ as a product of transpositions.

(c) Show that for any transpositions $\tau: \tau^{-1} = \tau$.

Answers:
(a) $(3, 7)(3, 4)(3, 6)(3, 1)$
(b) $(1, 4)(1, 2)(5, 6)(5, 8)(5, 10)(7, 9)$
(c) Se page A-13.

As previously noted, the decomposition of a permutation as a product of transposition is not unique. However:

THEOREM 2.34 No permutation can be expressed as both a product of an even number of transpositions and as a product of an odd number of transpositions.

PROOF: Chances are that you are familiar with the matrix space $M_{n \times n}$, along with the determinant function det: $M_{n \times n} \to \Re$. You may also recall that:

$$(*) \begin{cases} \text{If two rows of } A \in M_{n \times n} \text{ are interchanged, then, then} \\ \text{the determinant of the resulting matrix is } -\det(A). \end{cases}$$

(A brief development of the above result appears in Appendix B.)

Here is the identity matrix in M_4:

$$I = \begin{bmatrix} 1 & 0 & 0 & 0 \\ 0 & 1 & 0 & 0 \\ 0 & 0 & 1 & 0 \\ 0 & 0 & 0 & 1 \end{bmatrix} \begin{matrix} \leftarrow I_1 \\ \leftarrow I_2 \\ \leftarrow I_3 \\ \leftarrow I_4 \end{matrix}$$

And here is the matrix A obtained by the transposition (I_1, I_3):

$$A = \begin{bmatrix} 0 & 0 & 1 & 0 \\ 0 & 1 & 0 & 0 \\ 1 & 0 & 0 & 0 \\ 0 & 0 & 0 & 1 \end{bmatrix} \begin{matrix} \leftarrow \text{switched} \\ \text{first and} \\ \leftarrow \text{third row} \\ \text{of } I. \end{matrix}$$

At this point, rather then focusing on a permutation σ_i on the set $\{1, 2, \ldots, n\}$ of integer, we turn our attention to a permutation σ_I on the n rows $\{I_1, I_2, \ldots, I_n\}$ of the identity matrix $I \in M_{n \times n}$ (see margin). In this new environment, a transposition is the switching of two rows. Can $A \in M_{n \times n}$ be achieve by both an even number and an odd number of transpositions? No, for by (*), $\det(A)$ would have to equal both 1 and -1:

$$\det(A) = (-1)^{2k}\det(I) = 1 \text{ while } \det(A) = (-1)^{2k+1}\det(I) = -1$$

(Note: $\det(I) = 1$)

DEFINITION 2.12 A permutation is **even**, or **odd**, if it can be expressed as the product of an even, or odd, **Even and Odd Permutations** number of transpositions, respectively

---

CHECK YOUR UNDERSTANDING 2.28

(a) Show that the identity permutation $e \in S_n$ is even.

(b) Is the permutation $\begin{pmatrix} 1 & 2 & 3 & 4 & 5 & 6 & 7 & 8 & 9 & 10 \\ 2 & 4 & 3 & 1 & 10 & 5 & 9 & 6 & 7 & 8 \end{pmatrix}$ even or odd?

Answers: (a) See page A-13. (b) Even.

At this point we know that any symmetric group S_n can be partitioned into the set of even permutations and the set of odd permutations. Since the set of odd permutations does not contain the identity element [CYU 2.28(a)], it cannot be a subgroup of S_n. On the other hand:

DEFINITION 2.13 The **alternating group of degree** n is the **Alternating Group** subgroup A_n of even permutations of the symmetric group S_n.

Lest there be any doubt:

2.5 Symmetric Groups

THEOREM 2.35 For $n \geq 2$, the set A_n of even permutations is a subgroup of S_n of order $\dfrac{n!}{2}$.

PROOF: Since the identity permutation is even, $A_n \neq \emptyset$.

Closure: If σ and $\bar{\sigma}$ are even permutations, then each can be expressed as a product of an even number of transpositions, say:

$$\sigma = \tau_1 \tau_2 \cdots \tau_{2k} \text{ and } \bar{\sigma} = \bar{\tau}_1 \bar{\tau}_2 \cdots \bar{\tau}_{2h}$$

It follows that $\sigma\bar{\sigma}$ can be expressed as a product of $2(k+h)$ transpositions; namely: $\sigma\bar{\sigma} = \tau_1 \tau_2 \cdots \tau_{2k} \bar{\tau}_1 \bar{\tau}_2 \cdots \bar{\tau}_{2h}$

Inverse: If $\sigma = \tau_1 \tau_2 \cdots \tau_{2k}$, then σ^{-1} can also be expressed as a product of $2k$ transpositions; namely:

$$\sigma^{-1} \underset{\text{CYU 2.10, page 58}}{=} \tau_{2k}^{-1} \cdots \tau_2^{-1} \tau_1^{-1} \underset{\text{C YU 2.27(c).}}{=} \tau_{2k} \cdots \tau_2 \tau_1$$

Conclusion: A_n is a subgroup of S_n (Theorem 2.14, page 62).

Verifying that $|A_n| = \dfrac{n!}{2}$:

Let B_n denote the set of odd permutations in S_n. We show that the function $f: A_n \to B_n$ given by $f(\sigma) = (1,2)\sigma$ is a bijection.

One-to-one:

$$f(\sigma) = f(\bar{\sigma}) \Rightarrow (1,2)\sigma = (1,2)\bar{\sigma}$$
$$\Rightarrow (1,2)(1,2)\sigma = (1,2)(1,2)\bar{\sigma} \Rightarrow \sigma = \bar{\sigma}$$

Onto: For $\sigma \in B_n$, $(1,2)\sigma \in A_n$ and:

$$f[(1,2)\sigma] = (1,2)(1,2)\sigma = \sigma$$

We now know that A_n and B_n have the same number of elements. The fact that $A_n \cap B_n = \emptyset$ and that $A_n \cup B_n = S_n$ with $|S_n| = n!$ assures us that $|A_n| = \dfrac{n!}{2}$.

CHECK YOUR UNDERSTANDING 2.29

Determine A_3 utilizing the notation:

$$S_3 = \left\{ \begin{array}{l} e = \begin{pmatrix} 1 & 2 & 3 \\ 1 & 2 & 3 \end{pmatrix},\ \alpha_1 = \begin{pmatrix} 1 & 2 & 3 \\ 2 & 3 & 1 \end{pmatrix},\ \alpha_2 = \begin{pmatrix} 1 & 2 & 3 \\ 3 & 1 & 2 \end{pmatrix} \\ \alpha_3 = \begin{pmatrix} 1 & 2 & 3 \\ 1 & 3 & 2 \end{pmatrix},\ \alpha_4 = \begin{pmatrix} 1 & 2 & 3 \\ 3 & 2 & 1 \end{pmatrix},\ \alpha_5 = \begin{pmatrix} 1 & 2 & 3 \\ 2 & 1 & 3 \end{pmatrix} \end{array} \right\}$$

Answer: $A_3 = \langle \{e, \alpha_1, \alpha_2\}, \circ \rangle$

EXERCISES

Exercise 1-9. Express the given permutation as a product of disjoint cycles and also as a product of transpositions.

1. $\begin{pmatrix} 1 & 2 & 3 & 4 & 5 \\ 3 & 5 & 4 & 1 & 2 \end{pmatrix}$

2. $\begin{pmatrix} 1 & 2 & 3 & 4 & 5 \\ 5 & 2 & 4 & 1 & 3 \end{pmatrix}$

3. $\begin{pmatrix} 1 & 2 & 3 & 4 & 5 \\ 2 & 5 & 4 & 3 & 1 \end{pmatrix}^2$

4. $\begin{pmatrix} 1 & 2 & 3 & 4 & 5 & 6 \\ 6 & 5 & 4 & 1 & 2 & 3 \end{pmatrix}$

5. $\begin{pmatrix} 1 & 2 & 3 & 4 & 5 & 6 \\ 1 & 5 & 3 & 2 & 6 & 4 \end{pmatrix}$

6. $\begin{pmatrix} 1 & 2 & 3 & 4 & 5 & 6 \\ 5 & 4 & 3 & 1 & 2 & 6 \end{pmatrix}^2$

7. $\begin{pmatrix} 1 & 2 & 3 & 4 & 5 & 6 & 7 \\ 6 & 5 & 4 & 1 & 2 & 3 & 7 \end{pmatrix}$

8. $\begin{pmatrix} 1 & 2 & 3 & 4 & 5 & 6 & 7 \\ 5 & 7 & 4 & 1 & 2 & 3 & 6 \end{pmatrix}$

9. $\begin{pmatrix} 1 & 2 & 3 & 4 & 5 & 6 & 7 \\ 3 & 4 & 2 & 1 & 5 & 7 & 6 \end{pmatrix}^2$

Exercise 10-16. Find the order of the permutation in Exercise:

10. 1 11. 3 12. 4 13. 5 14. 7 15. 8 16. 9

Exercise 17-20. Solve for σ in the symmetric space S_5.

17. $(1, 3, 5)\sigma = (2, 4, 1)$

18. $(1, 3, 5)\sigma = (2, 4, 1)(4, 5)$

19. $(2, 4, 1)(4, 5)\sigma = (1, 3, 5)$

20. $(1, 3, 5)^2 \sigma = (2, 4, 1)^3$

21. Let a be an element of a group G. Show that the map $\lambda_a: G \to G$ given by $\lambda_a g = ag$ is a permutation on the set G.

22. Referring to Exercise 21, show that $H = \{\lambda_a | a \in G\}$ is a subgroup of S_G (the group of all permutations on G).

23. Prove that if σ, θ are disjoint cycles in S_n, then $\theta\sigma = \sigma\theta$.

24. Prove that there is no permutation σ such that $\sigma(1, 2)\sigma^{-1} = (1, 2, 3)$.

25. Prove that for any permutation σ and any transposition τ: $\sigma\tau\sigma^{-1}$ is a transposition.

26. Prove that if τ is a k-cycle, then $\sigma\tau\sigma^{-1}$ is also a k-cycle for any permutation σ.

27. Prove that there is a permutation σ such that $\sigma(1, 2, 3)\sigma^{-1} = (4, 5, 6)$.

28. Prove that every k-cycle in S_n has order k.

29. Prove that every cycle in S_n can be expressed as a product of transpositions.

30. Show that if σ is a cycle of odd length, then σ^2 is a cycle.

31. List all the elements in the alternating group of degree 4: A_4.

32. Let H be a subgroup of S_n. Prove that either all of the elements of H are even, or that exactly one-half the elements of H are even.

33. Express the k-cycle $(i_1, i_2, \ldots, i_k)$ as a product of k transpositions.

34. Let τ_1, τ_2 be transpositions with $\tau_1 \neq \tau_2$. Show that:

    (a) If τ_1 and τ_2 are disjoint, then $\tau_2 \tau_1$ can be expressed as the product of two 3-cycles.

    (b) If τ_1 and τ_2 are not disjoint, then $\tau_2 \tau_1$ can be expressed as a product of 3-cycles.

35. Prove that if $n \geq 3$, then every even permutation in S_n can be expressed as a product of 3-cycles. Suggestion: consider Exercise 34.

36. Let σ be a k-cycle. Show that $\sigma \in A_n$ if and only if $\sigma \tau \sigma^{-1} \in A_n$

37. Show that every $\sigma \in A_n$ with $n \geq 3$ is a product of 3-cycles.

	PROVE OR GIVE A COUNTEREXAMPLE	

38. The permutation equation $\sigma(1,2,3)\sigma^{-1} = (1,2,4)(5,6,7)$ has a solution.

39. The transposition $(1,2)$ in S_3 can be expressed as a product of 3-cycles.

40. The identity in S_n cannot be expressed as a product of three transpositions.

§6. NORMAL SUBGROUPS AND FACTOR GROUPS

An important recollection from page 67:

If H is a subgroup of a group G, then $a \sim b$ if $ab^{-1} \in H$ is an equivalence relation on G. Moreover, the equivalence class $[a]$ is the set:
$$Ha = \{ha | h \in H\} \text{ — a \textbf{right coset} of } H.$$

Let's switch from right to left, and replay the above development:

THEOREM 2.36 If H is a subgroup of a group G, then $a \sim b$ if $a^{-1}b \in H$ is an equivalence relation on G. Moreover, the equivalence class $[a]$ is the set:
$$aH = \{ah | h \in H\} \text{ — a \textbf{left coset} of } H.$$
If G is finite:
The number of elements in each aH equals $|H|$.

This proof mimics that of Lemma 2.2, page 67.

PROOF: $\sim$ **is reflexive:** $x \sim x$ since $x^{-1}x = e \in H$.

$\sim$ **is symmetric:** $a \sim b \Rightarrow a^{-1}b = h$ for some $h \in H$
$$\Rightarrow (a^{-1}b)^{-1} = h^{-1}$$
$$\Rightarrow b^{-1}a = h^{-1} \Rightarrow b \sim a \text{ since } h^{-1} \in H$$

$\sim$ **is transitive:** If $a \sim b$ and $b \sim c$, then:

$a^{-1}b \in H$ and $b^{-1}c \in H \Rightarrow (ab^{-1})(bc^{-1}) \in H$
$$\Rightarrow aec^{-1} \in H \Rightarrow ac^{-1} \in H \Rightarrow a \sim c$$

Having established the equivalence part of the theorem, we now verify that $[a] = \{ah | h \in H\}$:

$b \in [a] \Leftrightarrow b \sim a \Leftrightarrow a \sim b \Leftrightarrow a^{-1}b = h$ for some $h \in H$
$$\Leftrightarrow b = ah \Leftrightarrow b \in aH$$

As for the rest of the proof:

CHECK YOUR UNDERSTANDING 2.30

Let H be a subgroup of a finite group G. Show that each left coset aH contains $|H|$ elements.

Suggestion: Consider the function $f: H \to aH$.

Answer: See page B-14.

If H is a subgroup of an **abelian** group G then for any $a \in G$:
$$aH = \{ah | h \in H\} = \{ha | h \in H\} = Ha$$
(every left coset is also a right coset)
This need not be so if G is not abelian. A case in point:

$e = \begin{pmatrix} 1 & 2 & 3 \\ 1 & 2 & 3 \end{pmatrix}, \alpha_1 = \begin{pmatrix} 1 & 2 & 3 \\ 2 & 3 & 1 \end{pmatrix}$

$\alpha_2 = \begin{pmatrix} 1 & 2 & 3 \\ 3 & 1 & 2 \end{pmatrix}, \alpha_3 = \begin{pmatrix} 1 & 2 & 3 \\ 1 & 3 & 2 \end{pmatrix}$

$\alpha_4 = \begin{pmatrix} 1 & 2 & 3 \\ 3 & 2 & 1 \end{pmatrix}, \alpha_5 = \begin{pmatrix} 1 & 2 & 3 \\ 2 & 1 & 3 \end{pmatrix}$

EXAMPLE 2.10 Find the partition of S_3 (margin) into both the left and right cosets of the subgroup $H = \{e, \alpha_3\}$.

SOLUTION: Since $|S_3| = 6$ and $|H| = 2$, each partition is composed of 3 subsets, one of which is the subgroup H itself ($eH = He = H$). As for the rest of the story:

Left Cosets:

$eH = H = \{e, \alpha_3\}$

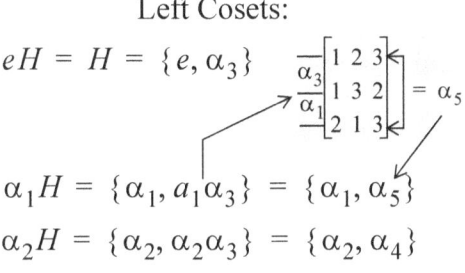

$\alpha_1 H = \{\alpha_1, \alpha_1\alpha_3\} = \{\alpha_1, \alpha_5\}$

$\alpha_2 H = \{\alpha_2, \alpha_2\alpha_3\} = \{\alpha_2, \alpha_4\}$

Left-Cosets Partition

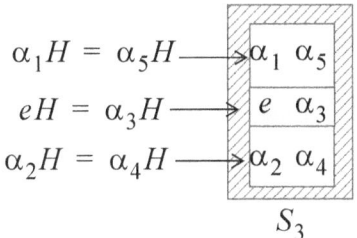

Right Cosets:

$He = H = \{e, \alpha_3\}$

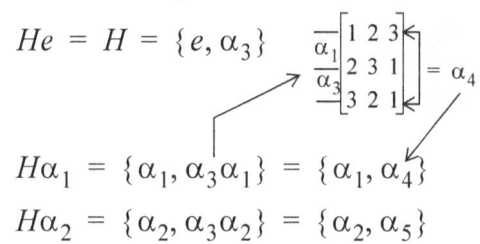

$H\alpha_1 = \{\alpha_1, \alpha_3\alpha_1\} = \{\alpha_1, \alpha_4\}$

$H\alpha_2 = \{\alpha_2, \alpha_3\alpha_2\} = \{\alpha_2, \alpha_5\}$

Right-Cosets Partition

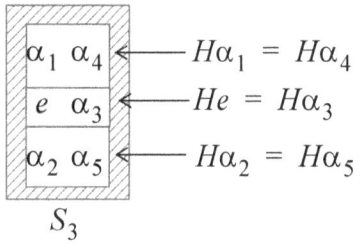

The above example illustrates the fact that a left cosets aH of a subgroup H of G need not equal the right coset Ha. Of particular importance are those subgroups for which left and right cosets are one and the same:

Clearly both G and $\{e\}$ are normal subgroups of any group G.

DEFINITION 2.14
NORMAL SUBGROUP

A subgroup N of a group G is said to be **normal in G** if for every $a \in G$:
$$aN = Na$$
(The symbol $N \triangleleft G$ is read: N is normal in G)

Index of N in G

If G is finite, then the number of cosets of N in G, namely $|G|/|H|$, is called the **index** of H in G.

THEOREM 2.37 Let H be a subgroup of a group G. The following are equivalent:

(i) $aH = Ha$ for every $a \in G$.
    (i.e: H is normal in G)

(ii) $aH \subseteq Ha$ for every $a \in G$.

(iii) $aha^{-1} \in H$ for every $h \in H$ and $a \in G$.

PROOF: $(i) \Rightarrow (ii)$: Clear.

$(ii) \Rightarrow (iii)$: Let $h \in H$ and $a \in G$ be given.

Since $aH \subseteq Ha$, $ah = \bar{h}a$ for some $\bar{h} \in H$. Consequently:
$$aha^{-1} = \bar{h} \in H$$

$(iii) \Rightarrow (i)$: We show that $Ha \subseteq aH$. A similar argument can be used to show that $aH \subseteq Ha$:

$$g \in aH \Rightarrow g = ah \text{ for } h \in H$$
$$\Rightarrow ga^{-1} = aha^{-1}$$
by (iii): $\Rightarrow ga^{-1} = \bar{h}$ for $\bar{h} \in H$
$$\Rightarrow g = \bar{h}a \Rightarrow g \in Ha$$

We showed, in Theorem 2.23, page 71, that homeomorphisms preserves subgroups. They fair nearly as well when it comes to normal subgroups. Specifically:

THEOREM 2.38 Let $\phi: G \to \bar{G}$ be a homomorphism.

(a) If N is normal in G, and if ϕ is onto, then $\phi(N)$ is normal in $\bar{G}$.

(b) If $\bar{N}$ is normal in $\bar{G}$, then $\phi^{-1}(\bar{N})$ is normal in G.

PROOF: (a) Assume that ϕ is onto and that N is normal in G.

We are to show that for any $\phi(n) \in \phi(N)$ and any $\bar{a} \in \bar{G}$, $\bar{a}\phi(n)\bar{a}^{-1} \in \phi(N)$. Let's do it:

Choose $a \in G$ such that $\phi(a) = \bar{a}$. Then:

$$\bar{a}\phi(n)\bar{a}^{-1} = \phi(a)\phi(n)[\phi(a)]^{-1}$$
Theorem 2.23(b), page 73: $= \phi(a)\phi(n)\phi(a^{-1})$
ϕ is a homomorphism: $= \phi(ana^{-1}) \underset{\underset{N \text{ is normal in } G}{\uparrow}}{\in} \phi(N)$

(b) Let $\bar{N}$ be normal in $\bar{G}$. We show that the subgroup
$$\phi^{-1}(\bar{N}) = \{n \in G | \phi(n) \in \bar{N}\} \text{ is normal in } G$$
by showing that for $a \in G$ and $n \in \phi^{-1}(\bar{N})$, $ana^{-1} \in \phi^{-1}(\bar{N})$:

$$\phi(ana^{-1}) = \phi(a)\phi(n)\phi(a^{-1}) \underset{\underset{\bar{N} \text{ is normal in } \bar{G}}{\uparrow}}{\in} \bar{N}$$

CHECK YOUR UNDERSTANDING 2.31

Give an example illustrating that a homomorphism $\phi: G \to \bar{G}$ that is not onto need not carry normal subgroups of G to normal subgroups of $\bar{G}$. (Suggestion: Consider Example 2.10.)

Answer: See page A-14.

2.6 Normal Subgroups and Factor Groups

Let's reconsider the left-coset partition of the subgroup $H = \{e, \alpha_3\}$ of Example 2.10. That partition, appearing on the right, broke the group S_3 into three disjoint pieces; each of which has "two names:"

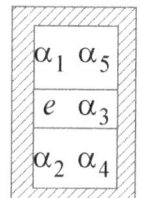

$$\alpha_1 H = \alpha_5 H, \ eH = \alpha_3 H, \text{ and } \alpha_2 H = \alpha_4 H$$

Can we impose a group structure on that partition? Here is a noble attempt:
$$(\alpha_i H)(\alpha_j H) = (\alpha_i \alpha_j)H$$

Yes, the above product certainly yields another left coset, but there is a fatal flaw — the "product" is simply **not defined**:

From Example 2-10:

$$\alpha_1 \alpha_2: \ \overset{\alpha_2}{\underset{\alpha_1}{\begin{bmatrix} 1 & 2 & 3 \\ 3 & 1 & 2 \\ 1 & 2 & 3 \end{bmatrix}}} = e$$

while:

$$\alpha_5 \alpha_4: \ \overset{\alpha_4}{\underset{\alpha_5}{\begin{bmatrix} 1 & 2 & 3 \\ 3 & 2 & 1 \\ 3 & 1 & 2 \end{bmatrix}}} = \alpha_2$$

$$\left. \begin{array}{l} \alpha_1 H = \alpha_5 H \\ \alpha_2 H = \alpha_4 H \end{array} \right\} \text{BUT:} \quad \begin{array}{l}(\alpha_1 \alpha_2)H = eH = H \\ \text{while} \\ (\alpha_5 \alpha_4)H = \alpha_2 H \end{array} \quad \text{(see margin)}$$

The above fatal flaw is averted whenever the coset-partition of a group G stems from a normal subgroup of G.

THEOREM 2.39 If $N \triangleleft G$ and
$$G/N = \{aN | a \in G\}$$
then G/N is a group under the operation
$$(aN)(bN) = (ab)N$$

> G/N is said to be the **factor group of G by N**,
> read: **G modulo H** or **G mod H**
> Factor groups are also said to be quotient groups.

PROOF: We first show that the operation $(aN)(bN) = (ab)N$ is well defined:

For $aN = \bar{a}N$ and $bN = \bar{b}N$ we need to establish the set equality $(ab)N = (\bar{a}\bar{b})N$. We show that $(ab)N \subseteq (\bar{a}\bar{b})N$ and leave it for you to verify that $(\bar{a}\bar{b})N \subseteq (ab)N$:

$$g \in (ab)N \Rightarrow g = abn \text{ for some } n \in N$$

Since $a \in \bar{a}N$ and $b \in \bar{b}N$: $\Rightarrow g = \bar{a}n_1 \bar{b} n_2 n$ for some $n_1, n_2 \in N$

$$\Rightarrow g = \bar{a}(n_1 \bar{b})n_3 \text{ where } n_3 = n_2 n$$

Since N is normal in G: $\Rightarrow g = \bar{a}(\bar{b}n_4)n_3$ for some $n_4 \in N$

$$\Rightarrow g = \bar{a}\bar{b}(n_4 n_3) \in (ab)N$$

Having legitimatized the operation $(aN)(bN) = (ab)N$, we now verify that, under that operation, the nonempty set G/N is a group (see Definition 2.1, page 41):

Closure: For every $aN, bN \in G/N$, $(aN)(bN) = (ab)N \in G/N$.

Associative: $(aNbN)(cN) = (ab)N(cN)$
$= [(ab)c]N = [a(bc)]N$
$= aN(bc)N = aN[bNcN]$

Identity: For every $aN \in G/N$, $aNeN = aN$.

Inverses: For every $aN \in G/N$, $aNa^{-1}N = aa^{-1}N = eN \ (=N)$.

> When confronted with the factor group G/N it is important that you keep in mind that you are dealing with a set of sets!
> In particular, the identity element in G/N is the set N itself, which may have many names:
> $$N = eN = aN \text{ for any } a \in N$$
> Similarly, the inverse of the element (set) aN is the element $a^{-1}N$, which may also have many names:
> $$a^{-1}N = (a^{-1}b)N \text{ for any } b \in N$$
> (after all, the set bN equals the set N for any $b \in N$)

THEOREM 2.40 Let N be normal in G. The natural projection map $\pi: G \to G/N$ given by $\pi(g) = gN$ is a homomorphism, and $\text{Ker}(\pi) = N$.

PROOF: For $g, \bar{g} \in G$: $\pi(g\bar{g}) = g\bar{g}N = gN\bar{g}N = \pi(g)\pi(\bar{g})$.

Moreover: $g \in \text{Ker}(\pi) \Leftrightarrow \pi(g) = gN = N \Leftrightarrow g \in N$

Recall that N is the identity in G/N

CHECK YOUR UNDERSTANDING 2.32

(a) Show that if G is a finite group and if $N \triangleleft G$, then:
$$|G/N| = \frac{|G|}{|N|}$$

(b) Let N be a normal subgroup of a cyclic group G. Prove that G/N is also cyclic.

Answer: See page A-14.

THE CENTER AND COMMUTATOR SUBGROUPS

Every group G contains two particularly important normal subgroups, the *center of G* and the *commutator subgroup of G*, where:

DEFINITION 2.15

CENTER OF G

The **center of G**, denoted by $Z(G)$, is the set of elements of G that commute with every element of G:
$$Z(G) = \{a \in G \mid ag = ga \ \forall g \in G\}$$

COMMUTATOR SUBGROUP OF G

The **commutator subgroup of G**, which we denote by $C(G)$, is the generated group:
$$C(G) = \langle aba^{-1}b^{-1} \mid a, b \in G \rangle$$

As advertised:

THEOREM 2.41 Both $Z(G)$ and $C(G)$ are normal subgroups of the group G.

PROOF: Turning to the center of G. Since $e \in Z(G)$, $\boldsymbol{Z(G) \neq \varnothing}$.

Closure: For $a, b \in Z(G)$, and any $g \in G$:
$$(ab)g = a(bg) \underset{b \in Z(G)}{=} a(gb) = (ag)b \underset{b \in Z(G)}{=} (ga)b = g(ab)$$

Inverses: For $a \in Z(g)$ and for any $g \in G$ (same as "for any $g^{-1} \in G$"):
$$ag = ga \Rightarrow (ag)^{-1} = (ga)^{-1} \Rightarrow g^{-1}a^{-1} = a^{-1}g^{-1}$$

Replacing g with g^{-1} in the above argument we conclude that a^{-1} commutes with every element of G.

Normal: Employing Theorem 2.36(ii) we show that for every $g \in G$
$$gZ(G) \subseteq Z(G):$$
$$x \in gZ(G) \Rightarrow x = ga \text{ for } a \in Z(G)$$
$$\Rightarrow x = ag \Rightarrow x \in Z(G)a$$

Now for $C(G)$. We already know that $C(G)$ is a subgroup of G (see Definition 2.8, page 65). As for the rest of the story:

Normal: For $x = aba^{-1}b^{-1}$ and any $c \in G$, let $x = cab$ and $y = a^{-1}b^{-1}c^{-1}$. We then have:
$$xyx^{-1}y^{-1} = (cab)(a^{-1}b^{-1}c^{-1})(b^{-1}a^{-1}c^{-1})(cba)$$
$$= (cab)(a^{-1}b^{-1}c^{-1}) = c(aba^{-1}b^{-1})c^{-1}$$
So: $c(aba^{-1}b^{-1})c^{-1} = xyx^{-1}y^{-1} \in C(G)$

Turning to Theorem 2.37(iii), and the Principle of Mathematical Induction, we now verify that $C(G)$ is normal in G.

I. If $x = (a_1b_1a_1^{-1}b_1^{-1})$, then, for any $c \in G$: $cxc^{-1} \in C(G)$.

II. Assume that for $n = k$ and any $c \in G$:
$x = (a_1b_1a_1^{-1}b_1^{-1})(a_2b_2a_2^{-1}b_2^{-1})\cdots(a_kb_ka_k^{-1}b_k^{-1}) \Rightarrow cxc^{-1} \in C(G)$

III. Then:

$$c[(a_1b_1a_1^{-1}b_1^{-1})(a_2b_2a_2^{-1}b_2^{-1})\cdots(a_kb_ka_k^{-1}b_k^{-1})(a_{k+1}b_{k+1}a_{k+1}^{-1}b_{k+1}^{-1})]c^{-1}$$
$$= c[(a_1b_1a_1^{-1}b_1^{-1})(a_2b_2a_2^{-1}b_2^{-1})\cdots(a_kb_ka_k^{-1}b_k^{-1})c^{-1}c(a_{k+1}b_{k+1}a_{k+1}^{-1}b_{k+1}^{-1})]c^{-1}$$
$$= \underbrace{[c(a_1b_1a_1^{-1}b_1^{-1})(a_2b_2a_2^{-1}b_2^{-1})\cdots(a_kb_ka_k^{-1}b_k^{-1})c^{-1}]}_{(*)}\underbrace{[c(a_{k+1}b_{k+1}a_{k+1}^{-1}b_{k+1}^{-1})c^{-1}]}_{(**)}$$

By II: $(*) \in C(G)$. By I: $(**) \in C(G)$
Consequently:
$$x = (a_1b_1a_1^{-1}b_1^{-1})\cdots(a_{k+1}b_{k+1}a_{k+1}^{-1}b_{k+1}^{-1}) \Rightarrow cxc^{-1} \in C(G)$$

THEOREM 2.42 For N normal in G, the factor group G/N is abelian if and only if $C(G) \subseteq N$.

PROOF: G/N is abelian if and only if for any $a, b \in G$:
$$aNbN = bNaN \Leftrightarrow abN = baN \Leftrightarrow (ba)^{-1}(ab) \in N$$
$$\Leftrightarrow a^{-1}b^{-1}ab \in N \Leftrightarrow C(G) \subseteq N$$

CHECK YOUR UNDERSTANDING 2.33

Let G be an abelian group. Show that $Z(G) = G$ and that $C(G) = \{e\}$.

Answer: See page A-14.

ISOMORPHISM THEOREMS

THEOREM 2.43
FIRST ISOMORPHISM THEOREM

If $\phi: G \to G'$ is a homomorphism, then $K = \text{Ker}(\phi)$ is normal in G and:
$$G/K \cong \phi(G)$$

PROOF: We utilize Theorem 2.37(iii) to establish the normality of K. For $k \in K$ and $g \in G$ we show that $gkg^{-1} \in K$; which is to say, that $\phi(gkg^{-1}) = e'$:
↑
the identity in G'

$$\phi(gkg^{-1}) = \phi(g)\phi(k)\phi(g^{-1})$$

Theorem 2.23(b), page 73: $= \phi(g)e'[\phi(g)]^{-1} = \phi(g)[\phi(g)]^{-1} = e'$

We complete the proof by showing that the function
$$\psi: G/K \to \phi(G) \text{ given by } \psi(gK) = \phi(g)$$
is an isomorphism. To begin with, we need to verify ψ is well defined:
$$aK = bK \Rightarrow ab^{-1} \in K \Rightarrow \phi(ab^{-1}) = e'$$
$$\Rightarrow \phi(a)\phi(b^{-1}) = e'$$
$$\Rightarrow \phi(a) = \phi(b) \Rightarrow \psi(aK) = \psi(bK)$$

ψ **is One-to-one:** We are to show that:
$$\psi(aK) = \psi(bK) \Rightarrow aK = bK$$
Which is to say: $\psi(aK) = \psi(bK) \Rightarrow ab^{-1} \in K$. Let's do it:
$$\psi(aK) = \psi(bK) \Rightarrow \phi(a) = \phi(b) \Rightarrow \phi(a)[\phi(b)]^{-1} = e'$$
$$\Rightarrow (\phi(a)\phi(b^{-1}) = e')$$
$$\Rightarrow \phi(ab^{-1}) = e' \Rightarrow ab^{-1} \in K$$

By II: (*) $\in C(G)$. By I: (**) $\in C(G)$
Consequently:
$$x = (a_1 b_1 a_1^{-1} b_1^{-1}) \cdots (a_{k+1} b_{k+1} a_{k+1}^{-1} b_{k+1}^{-1}) \Rightarrow cxc^{-1} \in C(G)$$

ψ is Onto: For given $\phi(g) \in \phi(G)$, $\psi(gK) = \phi(g)$.

ψ is a homomorphism:
$$\psi(aKbK) = \psi[(ab)K]$$
$$= \phi(ab) = \phi(a)\phi(b)$$
$$= \Psi(aK)\psi(bK)$$

EXAMPLE 2.11   Show that:
$$\langle Z_n, +_n \rangle \cong Z/(nZ)$$

SOLUTION: In Example 2.6, page 72, we showed that the function $\phi: \langle Z, + \rangle \to \langle Z_n, +_n \rangle$ given by $\phi(m) = r$ where $m = nq + r$ with $0 \leq r < n$ is a homomorphism.

While ϕ is not necessarily one-to-one it is certainly onto, as, for any $s \in Z_n$ $\phi(s) = s$. Applying Theorem 2.43, we then have:
$$Z_n \cong Z/K \text{ where } K = \text{Ker}(\phi).$$

Noting that:
$$\text{Ker}(\phi) = \{m | \phi(m) = 0\} = \{kn | k \in Z\} = nZ$$
$$\uparrow$$
$$\text{Example 2.4, page 62}$$

we conclude that: $Z_n \cong Z/(nZ)$

CHECK YOUR UNDERSTANDING 2.34

Answer: $G \cong S_n/A_n$

Represent the group $G = \{-1, 1\}$ (under standard integer multiplication) as a factor group of the symmetric group S_n.

Here are a couple more isomorphism theorems for your consideration:

THEOREM 2.44
SECOND ISOMORPHISM THEOREM

Let H be a subgroup of a group G, and N a normal subgroup of G. Then:
$$HN = \{hn | h \in H, n \in N\}$$
is a subgroup of G, $H \cap N$ is normal in G, and:
$$H/(H \cap N) \cong (HN)/N$$

PROOF: See Exercise 29.

THEOREM 2.45
THIRD ISOMORPHISM THEOREM

Let $\phi: G \to G'$ be an onto homomorphism with kernel K. If N' is normal in G', then:
$$N = \phi^{-1}(N') = \{a \in G | \phi(a) \in N'\}$$
is normal in G and:
$$G/N \cong G'/N'$$

PROOF: See Exercise 30.

CHECK YOUR UNDERSTANDING 2.35

Use Theorem 2.42 to verify that the isomorphism $G/N \cong G'/N'$ in Theorem 2.44 can also be expressed in the form:

$$G/N \cong (G/K)/(N/K)$$

("cancel" the K in the numerator and denominator)

Answer: See page A14.

2.6 Normal Subgroups and Factor Groups

EXERCISES

1-4. Determine if the give subgroup H is normal in the symmetric group S_3.

1. $H = \langle (1,2) \rangle$ 2. $H = \langle (1,2,3) \rangle$ 3. $H = \langle (1,3,2) \rangle$ 4. $H = A_3$

5. (a) Show that $Z \times Z / \langle (1,1) \rangle$ is an infinite cyclic group.

    (b) Show that $Z \times Z / \langle (2,2) \rangle$ is not a cyclic group

6. Show that if N_1 and N_2 are normal subgroups of G, then $N_1 \cap N_2$ is also normal in G.

7. Let N be a normal subgroup of G and let H be any subgroup of G. Show that $NH = \{nh \mid n \in N \text{ and } h \in H\}$ is a subgroup of G.

8. Let G be abelian and let H be a subgroup of G. Show that G/H is abelian.

9. Let G be cyclic and let H be a subgroup of G. Show that G/H is cyclic.

10. Let $\{N_\alpha\}_{\alpha \in A}$ be a collection of normal subgroup of G. Prove that $\bigcap_{\alpha \in A} N_\alpha$ is normal in G.

11. Show that if there are exactly 2 left (or right) cosets of a subgroup H of a group G, then $H \triangleleft G$.

12. Show that if a finite group G has exactly one subgroup H of a given order, then $H \triangleleft G$.

13. Show that if H is a finite subgroup of G and if H is the only subgroup of G with order $|H|$, then $H \triangleleft G$.

14. Let n be the index of the normal subgroup N in G. Show that $a^n \in N$ for every $a \in G$.

15. Let G be a group containing at least one subgroup of order n. Show that the intersection of all subgroups of order n in G is normal in G. Hint: first show that if a group H if of order n, then show that gHg^{-1} is also a subgroup of order n for all $g \in G$.

16. Show that the set of inner automorphisms of a group G is a normal subgroup of the group of all automorphisms of G. [see CYU 2.22(c), page 77]

17. Show that the set of all $g \in G$ for which the inner automorphism $i_g \colon G \to G$ is the identity map on G, is a normal subgroup of G. [see CYU 2.22(c), page 77]

18. Let N be a normal subgroup of G, and let $a, b, c, d \in G$ be such that $aN = cN$ and $bN = dN$. Show that $abN = cdN$.

19. Let G be a finite group of even order with n elements, and let H be a subgroup with $n/2$ elements. Prove that H *must* be normal. Suggestion: Consider the map $\phi \colon G \to \langle -(1, 1), \cdot \rangle$.

20. Let H and K be normal subgroups of G with $H \cap K = \{e\}$. Show that $hk = kh$ for all $h \in H$ and $k \in K$.

21. Let N be a normal subgroup of G such that G/N is cyclic. Show that G is cyclic.

22. Let G be a group. Show that any subgroup of $Z(G)$ is a normal subgroup of G.

23. Let G be a group. show that $C(a) = \{g \in G | ag = ga\}$ is a subgroup of G (called the **centralizer of** a).

24. Prove that the center of a group G is the intersection of all the centralizers in G; that is:
$$Z(G) = \bigcap_{a \in G} C(a) \text{ (See Exercise 22).}$$

25. Show that $a \in Z(G)$ if and only if $C(a) = G$. (See Exercise 22).

26. Find both the center and the commutator subgroup of S_3.

27. Let $\phi: G \to G'$ be an onto homomorphism with kernel K. Prove and if H' is a subgroup of G', and if $H = \phi^{-1}(H')$, then $H/K \cong H'$.

28. Verify that there is no subgroup of order 6 in the alternating group A_4. (Note that $|A_4| = 12$).

29. Sow that if N is not a normal subgroup of G, then the coset operation $(aN)(bN) = (ab)N$ is not well defined.

30. Prove Theorem 2.44.

31. Prove Theorem 2.45.

	PROVE OR GIVE A COUNTEREXAMPLE	

32. If $N \triangleleft G$ and if H is a subgroup of G, then $H \cap N \triangleleft G$.

33. If $H \triangleleft G$ and $K \triangleleft H$, then $K \triangleleft G$.

34. If $H \cap N \triangleleft G$ then either H or N must be normal in G.

35. Le $\phi: G \to G'$ be a homomorphism. If $N \triangleleft G$, then $\phi(N) \triangleleft G'$.

36. Le $\phi: G \to G'$ be a homomorphism. If $N \triangleleft G'$, then $\phi^{-1}(N') \triangleleft G$.

37. Le $\phi: G \to G'$ be an onto homomorphism. If $N \triangleleft G'$, then $\phi^{-1}(N') \triangleleft G$.

§7. DIRECT PRODUCTS

We begin by extending the Cartesian product definition of page 2:

DEFINITION 2.16

CARTESIAN PRODUCt

The **Cartesian Product** of n nonempty sets:
$$X_1, X_2, \ldots, X_n$$
is denoted by:
$$X_1 \times X_2 \times \cdots \times X_n \quad \left(\text{or: } \prod_{i=1}^{n} X_i\right)$$

and consists of all n-tuples:
$$(x_1, x_2, \ldots, x_n) \text{ where } x_i \in X_i \text{ for } 1 \le i \le n$$

In particular: $\Re \times \Re$ is the familiar Cartesian plane while $\Re \times \Re \times \Re$ is the Euclidean three-dimensional space.

Imposing a group structures we arrive at:

DEFINITION 2.17

DIRECT PRODUCT (EXTERNAL)

The (external) **Direct Product** of the n groups $G_1, G_2, \ldots, G_n$ is denoted by:
$$G_1 \times G_2 \times \cdots \times G_n \text{ or by } \prod_{i=1}^{n} G_i$$

consists of all ordered n-tuples
$$(a_1, a_2, \ldots, a_n) \text{ where } a_i \in G_i \text{ for } 1 \le i \le n$$

and where their multiplication is defined component-wise; that is:
$$(a_1, a_2, \ldots, a_n)(b_1, b_2, \ldots, b_n) = (a_1 b_1, a_2 b_2, \ldots, a_n b_n)$$

CHECK YOUR UNDERSTANDING 2.36

(a) Verify that $G_1 \times G_2 \times \cdots \times G_n$ is a group.

(b) Prove that the group $G_1 \times G_2 \times \cdots \times G_n$ is abelian if and only if each G_i is abelian.

Answer: See page A-15.

EXAMPLE 2.12 (a) Verify that $Z_2 \times Z_3$ is cyclic.

(b) Verify that $Z_2 \times Z_3 \times Z_4$ is not cyclic.

SOLUTION: (a) We know that $|Z_2 \times Z_3| = 2 \cdot 3 = 6$. Using the sum notation in the abelian group, we simply observe that the element $(1, 1)$ has order 6:

$$2(1, 1) = (0, 2)$$
$$3(1, 1) = (1, 1) + 2(1, 1) = (1, 1) + (0, 2) = (1, 0)$$
$$4(1, 1) = (1, 1) + 3(1, 1) = (1, 1) + (1, 0) = (0, 1)$$
$$5(1, 1) = (1, 1) + 4(1, 1) = (1, 1) + (0, 1) = (1, 2)$$
$$6(5)(1, 1) = (1, 1) + 5(1, 1) = (1, 1) + (1, 2) = (0, 0) \text{ Ah!}$$

(b) We show that the group $Z_2 \times Z_3 \times Z_4$, which is of order 24, contains no element of order greater than 12:

Let $(a, b, c) \in Z_2 \times Z_3 \times Z_4$. Since:

$$12a = 0, 12b = 0, 12c = 0 \text{ in the groups } Z_2, Z_3, Z_4,$$

respectively: $12(a, b, c) = (0, 0, 0)$.

In the above argument, 12 is the smallest positive integer that is divisible by 2, 3, and 4. In general:

DEFINITION 2.18
LEAST COMMON MULTIPLE
The **least common multiple** of nonzero integers $a_1, a_2, ..., a_n$, written $\text{lcm}(a_1, a_2, ..., a_n)$ is the smallest positive integer that is a multiple of each a_i; i.e. is divisible by each a_i.

THEOREM 2.46

Let $(a_1, a_2, ..., a_n) \in \prod_{i=1}^{n} G_i$. If the order of a_i in G_i is r_i, then the order of $(a_1, a_2, ..., a_n)$ in $G_1 \times G_2 \times \cdots \times G_n$ is $\text{lcm}(r_1, r_2, ..., r_n)$.

PROOF: Let $M = \text{lcm}(r_1, r_2, ..., r_n)$. Since each $r_i | M$:

$$(a_1, a_2, ..., a_n)^M = (e_1, e_2, ..., e_n)$$

Moreover, for any positive integer $0 < m < M$:

$$(a_1, a_2, ..., a_n)^m = (a_1^m, a_2^m, ..., a_n^m) \neq (e_1, e_2, ..., e_n)$$

Why is that so? Because since M is the smallest positive integer divisible by each r_i, some $a_i^m \neq e_i$.

EXAMPLE 2.13 Find the order of $(1, 5, 4)$ in $Z_2 \times Z_6 \times Z_{30}$.

SOLUTION: Let r_1, r_2, r_3 denote the order of 1 in Z_2, the order of 5 in Z_6, and the order of 4 in Z_{30}, respectively.

Employing CYU 2.11(c), page 59 (margin) we find that:

$$r_1 = 2, \ r_2 = 6, \ r_3 = 15$$

For $m \in Z_n$:
$$o(m) = \frac{n}{gcd(m, n)}$$

All that remains is to calculate the least common multiple of the above orders:
$$o(1, 5, r) = \text{lcm}(2, 6, 15) = 30$$

2, $2 \cdot 3$, $3 \cdot 5$: need one 2, one 3, and one 5: $2 \cdot 3 \cdot 5$

CHECK YOUR UNDERSTANDING 2.37

Answer: 4

Determine the order of $(3, 3, 4)$ in $Z_6 \times Z_4 \times Z_{16}$.

THEOREM 2.47 The group $Z_n \times Z_m$ is cyclic and isomorphic to Z_{nm} if and only if n and m are relatively prime.

PROOF: Assume that n and m are relatively prime. Theorem 2.42 tels us that $o(1, 1) = nm$; which is to say:
$$Z_n \times Z_m = \langle (1, 1) \rangle$$
Since $|Z_n \times Z_m| = nm$:
$$Z_n \times Z_m \cong \langle Z_{nm}, +_n \rangle \quad \text{[see Theorem 2.26(a), page 77]}$$

As for the converse, assume that $gcd(n, m) = d > 1$. Noting that $\frac{nm}{d}$ is divisible by both n and m, we find that, for any $(a, b) \in Z_n \times Z_m$: $\frac{nm}{d}(a, b) = (0, 0)$. Since no element of $Z_n \times Z_m$ has order greater than $\frac{nm}{d}$, $Z_n \times Z_m$ is not cyclic.

CHECK YOUR UNDERSTANDING 2.38

Prove: The group $Z_{n_1} \times Z_{n_2} \times \cdots \times Z_{n_s}$ is cyclic and isomorphic to $Z_{n_1 n_2 \cdots n_s}$ if and only all pair of the numbers $n_1, n_2, \ldots, n_s$ are relatively prime.

Answer: See page A-15.

INTERNAL DIRECT PRODUCT

On the surface, the following definition appears to be far removed from Definition 2.17:

DEFINITION 2.19

DIRECT PRODUCT (INTERNAL)

A group G is said to be the (internal) **direct product** of n normal subgroups
$$N_1, N_2, \ldots, N_n$$
if every $g \in G$ has a unique representation of the form
$$g = a_1 a_2 \ldots a_n$$
where each $a_i \in N_i$ for $1 \leq i \leq n$.

> **CHECK YOUR UNDERSTANDING 2.39**
>
> Show that if G is the internal direct product of two normal subgroups H and K, then $H \cap K = \{e\}$.

Answer: See page A-16.

Appearances aside, the internal and external direct product concepts are "algebraically equivalent," in that every internal product space is isomorphic to an external product space, and every external product space is isomorphic to an internal product space.

Taking the easy way out, we will content ourselves by establishing the above claim in the special case when the group G is the internal direct product of just two normal subgroups:

THEOREM 2.48 (a) If G is the internal direct product of the normal subgroups H and K, then:
$$HK \cong H \times K$$
(b) If $G = G_1 \times G_2$, then there exist normal subgroups N_1 and N_2 in G such that:
$$G_1 \times G_2 \cong N_1 N_2$$

PROOF: (a) We first show that for all $h \in H$ and $k \in K$
$$hk = kh \quad (*):$$
Since $h^{-1} \in H$ and $H \triangleleft G$: $kh^{-1}k^{-1} \in H$. So: $h(kh^{-1}k^{-1}) \in H$.
Since $K \triangleleft G$: $hkh^{-1} \in K$. So: $(hkh^{-1})k^{-1} \in K$.
Since, by CYU 2.39, $hkh^{-1}k^{-1} \in H \cap K = \{e\}$: $hk = kh$.

Turning to the external product $H \times K$ of the two groups H and K, we now show that the function $\phi: H \times K \to HK$ given by $\phi(h, k) = hk$ is an isomorphism:

One to one: $\phi(h_1, k_1) = \phi(h_2, k_2) \Rightarrow h_1 k_1 = h_2 k_2$
$$\Rightarrow h_1 = h_2 \text{ and } k_1 = k_2$$
$$\Rightarrow (h_1, k_1) = (h_2, k_2)$$

Onto: Clear.

Homomorphism: For $(h_1, k_1), (h_2, k_2) \in H \times K$:
$$\phi[(h_1, k_1)(h_2, k_2)] = \phi(h_1 h_2, k_1 k_2) = h_1 h_2 k_1 k_2$$
$$\text{By } (*): = h_1 k_1 h_2 k_2$$
$$= \phi(h_1, k_1)\phi(h_2, k_2)$$

(b) Let $G = G_1 \times G_2$. It is easy to see that:
$$N_1 = \{(e_1, g) | g \in G\} \text{ and } N_2 = \{(g, e_2) | g \in G\}$$
are normal subgroups of G with $N_1 \cap N_2 = \{e\} \leftarrow (e_1, e_2)$, and that $G = N_1 N_2$. That being the case, the identity map itself is an isomorphism from the external direct product $G = G_1 \times G_2$ to the internal direct product $G = N_1 N_2$.

THE FUNDAMENTAL THEOREM OF FINITELY GENERATED ABELIAN GROUPS

And here it be, presented without proof:

THEOREM 2.49 Every finitely generated abelian group is isomorphic to a direct product of cyclic groups of the form:

$$Z_{p_1^{r_1}} \times Z_{p_2^{r_2}} \times \cdots \times Z_{p_n^{r_n}} \times Z^m$$

where the p_i are primes, **not** necessarily distinct, and where the r_i and m are positive integers.

Moreover, the direct product is unique, up to order.

For example:

$$G = Z_2 \times Z_8 \times Z_9 \times Z \times Z \quad (*)$$

is a finitely generated abelian group, and here is a particularly nice choice for its generators:

$(1, 0, 0, 0, 0), (0, 1, 0, 0, 0), (0, 0, 1, 0, 0), (0, 0, 0, 1, 0), (0, 0, 0, 0, 1)$
↑ could have chosen any $1 \le i \le 7$ ↑ could also have chosen 2, 4, 5, 7, or 8 ↑ could have chosen any non-zrto integer

While an abelian group generated by an element of order 2, one of order 8, another of order 9, and a couple of generators of infinite order need not consist of 5-tuples, it is nonetheless isomorphic to (*).

EXAMPLE 2.14 Find all abelian groups of order 600 (up to isomorphism).

SOLUTION: Any finite abelian group G is surely finitely generated (the elements of G itself generate G).
Employing Theorem 2.45 to:

$$600 = 2^3 \cdot 3 \cdot 5^2$$

we arrive at the following six possibilities (see margin):

$G_1 = Z_2 \times Z_2 \times Z_2 \times Z_3 \times Z_5 \times Z_5$

$G_2 = Z_2 \times Z_2 \times Z_2 \times Z_3 \times Z_{25}$

$G_3 = Z_4 \times Z_2 \times Z_3 \times Z_5 \times Z_5$

$G_4 = Z_8 \times Z_3 \times Z_5 \times Z_5$

$G_5 = Z_4 \times Z_2 \times Z_3 \times Z_{25}$

$G_6 = Z_8 \times Z_3 \times Z_{25}$

FUNDAMENTAL COUNTING PRINCIPLE:
If each of n choices is followed by m choices, then the total number of choices is given by $n \cdot m$.
There are three choices for the number of 2's in the direct product, a choice of one for the number of 3's, and a choice of two for the number of 5's.
Total number of choices:
$3 \cdot 1 \cdot 2 = 6$

It can be shown that none of the above groups is isomorphic to any of the rest. For example, since G_3 contains an element of order 4 while G_1 does not: $G_1 \not\cong G_3$ (see Exercise 36, page 82).

CHECK YOUR UNDERSTANDING 2.40

Referring to the above example, show that $G_3 \not\cong G_6$.

Answer: See page A-15.

> The alternating group A_4, of order 12, has **no** subgroup of order 6. Yes, but A_4 is **not** an abelian group.

Lagrange's Theorem assures us that the order of any subgroup H of a finite group G must divide the order of G. In the event that G is abelian, the converse also holds:

THEOREM 2.50 If m divides the order of an abelian group G, then G has a subgroup of order m.

PROOF: Theorem 2.45 enables us to express G in the form:

$$Z_{p_1^{r_1}} \times Z_{p_2^{r_2}} \times \cdots \times Z_{p_n^{r_n}}$$

Since m divides the order of G:

$$m = p_1^{s_1} p_2^{s_2} \ldots p_n^{s_n}, \text{ where } 0 \leq s_i \leq r_i.$$

By CYU 2.15, page 64:

$$o\langle p_i^{r_i - s_i}\rangle = \frac{p_i^{r_i}}{gcd(p_i^{r_i}, p_i^{r_i - s_i})} = p_i^{r_i - (r_i - s_i)} = p^{s_i}$$

It follows that:

$$\langle p_1^{r_1 - s_1}\rangle \times \langle p_2^{r_2 - s_2}\rangle \times \ldots \times \langle p_n^{r_n - s_n}\rangle$$

is a subgroup of G of order m.

EXERCISES

Exercise 1-6. Find the order of the given element if the give group.

1. $(2, 3)$ in $Z_4 \times Z_9$
2. $(2, 3)$ in $Z_5 \times Z_{12}$
3. $(2, 2, 8)$ in $Z_4 \times Z_3 \times Z_{12}$
4. $(2, 2, 8)$ in $Z_4 \times Z_6 \times Z_{11}$
5. $\left(2, \begin{pmatrix} 1 & 2 & 3 \\ 2 & 1 & 3 \end{pmatrix}\right)$ in $Z_4 \times S_3$
6. $\left(3, \begin{pmatrix} 1 & 2 & 3 \\ 2 & 3 & 1 \end{pmatrix}\right)$ in $Z_4 \times S_3$

Exercise 7-10. Find the order of each element if the given group.

7. $Z_2 \times Z_3$
8. $Z_2 \times Z_4$
9. $Z_2 \times Z_2 \times S_2$
10. $Z_3 \times S_2$

Exercise 11-14. Find all proper subgroups of the given group.

11. $Z_2 \times Z_3$
12. $Z_2 \times Z_4$
13. $Z_2 \times Z_2 \times S_2$
14. $Z_3 \times S_2$

Exercise 15-18. Find all abelian groups G of the give order (up to isomorphism).

15. $o(G) = 36$
16. $o(G) = 100$
17. $o(G) = 180$
18. $o(G) = 243$

19. Determine the number of elements of order 6 in $Z_6 \times Z_9$.

20. Determine the number of elements of order 7 in $Z_{49} \times Z_7$.

21. Show that the Klein 4-group V (Figure 2.1, page 43) is isomorphic to $Z_2 \times Z_2$.

22. Show that $(Z \times Z) / \langle (1, 1) \rangle \cong Z$.

23. Show that $(Z \times Z \times Z) / \langle (1, 1, 1) \rangle \cong Z \times Z$.

24. Use the Principle of Mathematica Induction to show that for finite groups $G_1, G_2, \ldots, G_n$:
$$|G_1 \times G_2 \times \cdots \times G_n| = |G_1||G_2|\cdots|G_n|$$

25. Let G_1 and G_2 be groups. Show that $G_1 \times G_2 \cong G_2 \times G_1$.

26. Let G_1 and G_2 be groups. Show that $Z(G_1 \times G_2) \cong Z(G_1) \times Z(G_2)$ (see Definition 2.15, page 96).

27. Let G_1 and G_2 be groups. Show that $\{e_1\} \times G_2 \triangleleft G_1 \times G_2$ and that:
$$(G_1 \times G_2)/(\{e_1\} \times G_2) \cong G_2$$

28. Let $H \triangleleft G_1$ and $K \triangleleft G_2$. Show that $H \times K \triangleleft G_1 \times G_2$ and that:
$$(G_1 \times G_2)/(H \times K) \cong G_1/H \times G_2/K$$

29. Let G_1 and G_2 be groups. Show that the order of $(a, b) \in G_1 \times G_2$ is the leas common multiple of $o(g)$ and $o(h)$.

30. Prove that the order of an element in a direct product of a finite number of finite groups $\{G_i\}_{i=1}^{n}$ is the least common multiple of the orders of its components:
$$o(g_1, g_2, \ldots, g_n) = lcm[o(g_1), o(g_2), \ldots, o(g_n)]$$

31. Let G be a group and $K = \{(g, g) | (g \in G)\} \subseteq G \times G$. Prove that
    (a) $K \cong G$
    (b) $K \triangleleft G \times G$ if and only if G is abelian.

32. Let $G = G_1 \times G_2 \times \cdots \times G_n$ be a direct product of groups. Show that the **projection map** $\pi_i: G \to G_i$ given by $\pi_i(g_1, g_2, \ldots, g_i, \ldots, g_n) = g_i$ is a homomorphism

	PROVE OR GIVE A COUNTEREXAMPLE	

33. The groups $Z_2 \times Z_{12}$ and $Z_4 \times Z_6$ are isomorphic.

34. The groups $Z_2 \times Z_4 \times Z_8$ and $Z_8 \times Z_8$ are isomorphic.

35. The groups $Z_2 \times Z_3 \times Z_8$ and $Z_3 \times Z_4 \times Z_4$ are isomorphic.

36. Let G, H, K denote groups. If $G \times K \cong H \times H$, then $G \cong H$.

Part 3
From Rings To Fields

§1. DEFINITIONS AND EXAMPLES

The familiar set of integers can boast or two operators: addition and multiplication. Though The integers under addition turns out to be an abelian group, the multiplication operator does not fair as well: (5, for example, has no multiplicative inverse).

Multiplication is, however, an associative operator:
$$a(bc) = (ab)c \ \forall a, b, c \in Z$$
and it plays well with addition:
$$a(b+c) = ab + ac \ \forall a, b, c \in Z$$

Just as the integers under addition directed us, in part, to the definition of a group ("in part," as a group need not be abelian), so then do the integers under addition and multiplication direct us, in part, to the definition of a ring ("in part," as a ring need not have a multiplicative identity). Specifically:

DEFINITION 3.1
RING

A **ring** $\langle R, +, \cdot \rangle$ (or simply R) is a set R together with two binary operators, called addition and multiplication; for which:

Group Axiom: 1. $\langle R, + \rangle$ is an **abelian** group.

Associativity Axiom: 2. For all $a, b, c \in R$: $a \cdot (b \cdot c) = (a \cdot b) \cdot c$
(multiplicative)

Distributive Axioms: 3. For all $a, b, c \in R$:
$$a \cdot (b+c) = a \cdot b + a \cdot c$$
$$(a+b) \cdot c = a \cdot c + b \cdot c$$

From an axiomatic point of view, multiplication takes a back seat to addition. Its only obligation, apart from closure and the associative axiom, is to cooperate with addition via the left and right distributive property of Axiom 3.

The set Z of integers under standard addition and multiplication is a ring. The same can be said for the set Q or rational numbers, and the set $\mathfrak{R}$ of reals.

CHECK YOUR UNDERSTANDING 3.1

(a) Does there exist an operator "$*$" on the permutation group $S_3 = \langle S_3, \circ \rangle$ for which $\langle S_3, \circ, * \rangle$ is a ring?

(b) Let $\langle G, + \rangle$ be an abelian group. Show that there exists an operator "$*$" on G for which $\langle G, +, * \rangle$ is a ring.

Answer: (a) No
(b) See page A-16.

EXAMPLE 3.1 Let R_1 and R_2 be rings. Prove that the group $\langle R_1 \times R_2, +, \cdot \rangle$ where

$$(a, b) + (c, d) = (a + c, b + d)$$
$$(a, b)(c, d) = (ac, bd)$$

is a ring.

Noe that addition and multiplication in $R_1 \times R_2$ are both being defined in terms of their corresponding established operations in R_1 and R_2.

SOLUTION: Appealing directly to Definition 2.1, page 41, we first show that $\langle R_1 \times R_2, + \rangle$ is an abelian group:

Associative:

For any $(a_1, b_1), (a_2, b_2), (a_3, b_3) \in R_1 \times R_2$:

$$(a_1, b_1) + [(a_2, b_2) + (a_3, b_3)] = (a_1, b_1) + (a_2 + a_3, b_2 + b_3)$$
$$= [a_1 + (a_2 + a_3), b_1 + (b_2 + b_3)]$$
$$= [(a_1 + a_2) + a_3, (b_1 + b_2) + b_3] = [(a_1, b_1) + (a_2, b_2)] + (a_3, b_3)$$

Identity: For any given $(a, b) \in R_1 \times R_2$:

$$(a, b) + (0, 0) = (a + 0, b + 0) = (a, b)$$

Inverses: For any given $(a, b) \in R_1 \times R_2$:

$$(a, b) + (-a, -b) = (a - a, b - b) = (0, 0)$$

Noting that for any $(a_1, b_1), (a_2, b_2) \in R_1 \times R_2$:

$$(a_1, b_1) + (a_2, b_2) = (a_1 + a_2, b_1 + b_2) = (a_2 + a_1, b_2 + b_1) = (a_2, b_2) + (a_1, b_1)$$

we conclude that $\langle R_1 \times R_2, + \rangle$ is an abelian group.

Moving on to the multiplicative axioms of Definition 3.1:

Associative: For any $(a_1, b_1), (a_2, b_2), (a_3, b_3) \in R_1 \times R_2$

$$(a_1, b_1)[(a_2, b_2)(a_3, b_3)] = (a_1, b_1)(a_2 a_3, b_2 b_3)$$
$$= [a_1(a_2 a_3), b_1(b_2 b_3)]$$
$$= [(a_1 a_2)a_3, (b_1 b_2)b_3] = [(a_1, b_1)(a_2, b_2)](a_3, b_3)$$

Distributive: For any $(a_1, b_1), (a_2, b_2), (a_3, b_3) \in R_1 \times R_2$

$$(a_1, b_1)[(a_2, b_2) + (a_3, b_3)] = (a_1, b_1)(a_2 + a_3, b_2 + b_3)$$
$$= [a_1(a_2 + a_3), b_1(b_2 + b_3)]$$
$$= (a_1 a_2 + a_1 a_3, b_1 b_2 + b b_3)$$
$$= (a_1 a_2, b_1 b_2) + (a_1 a_3, b_1 b_3) = (a_1, b_1)(a_2, b_2) + (a_1, b_1)(a_3, b_3)$$

In a similar fashion once can show that:

$$[(a_1, b_1) + (a_2, b_2)](a_3, b_3) = (a_1, b_1)(a_3, b_3) + (a_2, b_2)(a_3, b_3)$$

CHECK YOUR UNDERSTANDING 3.2

Sow that nZ, under standard addition and multiplication, is a ring.

Answer: See page A-16

3.1 Definitions and Examples 113

THEOREM 3.1 Let a and b be elements of a ring R. Then:

(a) $a0 = 0a = 0$

(b) $a(-b) = (-a)b = -ab$

(c) $(-a)(-b) = ab$

(d) $n(ab) = (na)b = a(nb)$ for any integer n.
(see margin)

> There is only one product taking place in $n(ab)$; namely the ab. The n is not involved in a product— it represents a sum. For example:
> $3(ab) = ab + ab + ab$

PROOF: (a) $a0 = a(0+0) = a0 + a0$

$a0 - a0 = a0$

$0 = a0$

(b) Since $a(-b) + ab = a(-b+b) = a0 = 0: a(-b) = -ab$.

Since $(-a)b + ab = (-a+a)b = 0b = 0: (-a)b = -ab$.

Since $a(-b) = -ab$ and $(-a)b = -ab: a(-b) = (-a)b$.

> Given your arithmetic evolution, you may be thinking along these lines:
> $(-a)(-b) = (-1a)(-1b) = ab$
> Tisk. For one thing, the ring R need not even have a unity. That $-a$, for example, is the additive inverse of a. That being the case: $-(-a) = a$

(c) $(-a)(-b) \underset{\text{by (b)}}{=} [-(-a)]b = ab$ (see margin).

As for (d):

CHECK YOUR UNDERSTANDING 3.3

Let a and b be elements of a ring R. Show that for every $n \in Z$:
$n(ab) = (na)b = a(nb)$

Answer: See page A-16

While Definition 3.1 stipulates that addition is a commutative operator tin a ring $\langle R, +, \cdot \rangle$, no such attribute is imposed on the product operator. Moreover, while every ring contains the additive identity "0", a ring need not contain a **multiplication identity** "1" (see margin).

> An element $a \in R$ distinct from 0 is a multiplicative identity (or unity) if for every $b \in R: ab = ba = b$.

Bringing us to:

DEFINITION 3.2
Commutative Ring   A ring $\langle R, +, \cdot \rangle$ is said to be **commutative** if $ab = ba$ for every $a, b \in R$.

Ring with Unity   A ring with a multiplicative identity (or **unity**) is said to be a **ring with unity**.

For any $n > 1$, the commutative ring nZ of CYU 3.2 is an example of a ring that does not have a unity. Here is an example of a ring that is not commutative:

114   Part 3   From Rings To Fields

EXAMPLE 3.2 Show that the set of two-by-two matrices:

$$M_{2\times 2} = \left\{ \begin{bmatrix} a & b \\ c & d \end{bmatrix} \middle| a, b, c, d \in \Re \right\}$$

with addition and multiplication given by:

$$\begin{bmatrix} a & b \\ c & d \end{bmatrix} + \begin{bmatrix} \bar{a} & \bar{b} \\ \bar{c} & \bar{d} \end{bmatrix} = \begin{bmatrix} a+\bar{a} & b+\bar{b} \\ c+\bar{c} & d+\bar{d} \end{bmatrix}$$

Chances are that you are already familiar with the matrix space $M_{n\times n}$ which possesses both an additive and multiplicative structure. If so, then you already know that, for any $n \geq 2$, $M_{n\times n}$ is a non-commutative ring.

and

$$\begin{bmatrix} a & b \\ c & d \end{bmatrix} \begin{bmatrix} \bar{a} & \bar{b} \\ \bar{c} & \bar{d} \end{bmatrix} = \begin{bmatrix} a\bar{a}+b\bar{c} & a\bar{b}+b\bar{d} \\ c\bar{a}+d\bar{c} & c\bar{b}+d\bar{d} \end{bmatrix}$$

is a non-commutative ring.

SOLUTION: In CYU 3.4 below you are invited to show that $M_{2\times 2} = \langle M_{2\times 2}, +, \cdot \rangle$ is a ring.

Is it a commutative ring? No:

$$\begin{bmatrix} 1 & 0 \\ 0 & 0 \end{bmatrix}\begin{bmatrix} 0 & 1 \\ 0 & 0 \end{bmatrix} = \begin{bmatrix} 0 & 1 \\ 0 & 0 \end{bmatrix} \text{ while } \begin{bmatrix} 0 & 1 \\ 0 & 0 \end{bmatrix}\begin{bmatrix} 1 & 0 \\ 0 & 0 \end{bmatrix} = \begin{bmatrix} 0 & 0 \\ 0 & 0 \end{bmatrix}$$

CHECK YOUR UNDERSTANDING 3.4

(a) Verify that $M_{2\times 2} = \langle M_{2\times 2}, +, \cdot \rangle$ is a ring with unity.

Answer: See page A-17.

(b) Prove that if a ring contains a unity, then that unity is unique.

You need to distinguish between "unity" and "unit."
unity: Multiplicative identity.
unit: An element that has a multiplicative inverse.

DEFINITION 3.3 Let R be a ring with unity 1. An element
Unit $a \in R$ is a **unit** if there exists $b \in R$ such that $ab = ba = 1$.

The element b is called the inverse of a and is denoted by a^{-1}.

EXAMPLE 3.3 (a) Show that the ring $M_{2\times 2}$ of Example 3.2 has a unity.

(b) Show that $\begin{bmatrix} -5 & 2 \\ 9 & -4 \end{bmatrix}$ is a unit, and that $\begin{bmatrix} 2 & 3 \\ -4 & -6 \end{bmatrix}$ is not a unit in $M_{2\times 2}$.

SOLUTION: (a) $\begin{bmatrix} 1 & 0 \\ 0 & 1 \end{bmatrix}$ is the unity in $M_{2\times 2}$:

$$\begin{bmatrix} 1 & 0 \\ 0 & 1 \end{bmatrix}\begin{bmatrix} a & b \\ c & d \end{bmatrix} = \begin{bmatrix} a & b \\ c & d \end{bmatrix}\begin{bmatrix} 1 & 0 \\ 0 & 1 \end{bmatrix} = \begin{bmatrix} a & b \\ c & d \end{bmatrix}$$

A linear algebra approach:

Since $\det\begin{bmatrix} -5 & 2 \\ 9 & -4 \end{bmatrix} \neq 0$,

$\begin{bmatrix} -5 & 2 \\ 9 & -4 \end{bmatrix}$ is invertible.

(b) Does there exist a matrix $\begin{bmatrix} a & b \\ c & d \end{bmatrix}$ for which:

$$\begin{bmatrix} -5 & 2 \\ 9 & -4 \end{bmatrix}\begin{bmatrix} a & b \\ c & d \end{bmatrix} = \begin{bmatrix} a & b \\ c & d \end{bmatrix}\begin{bmatrix} -5 & 2 \\ 9 & -4 \end{bmatrix} = \begin{bmatrix} 1 & 0 \\ 0 & 1 \end{bmatrix}?$$

Let's see:

$$\begin{bmatrix} -5 & 2 \\ 9 & -4 \end{bmatrix}\begin{bmatrix} a & b \\ c & d \end{bmatrix} = \begin{bmatrix} 1 & 0 \\ 0 & 1 \end{bmatrix} \Leftrightarrow \begin{bmatrix} 2a+3c & 2b+3d \\ -4a-6c & -4b-6d \end{bmatrix} = \begin{bmatrix} 1 & 0 \\ 0 & 1 \end{bmatrix}$$

$$\Leftrightarrow \left.\begin{array}{r} 2a+3c = 1 \\ 2b+3d = 0 \\ -4a-6c = 0 \\ -ab-6d = 1 \end{array}\right\}$$

If you take the time to solve the above system of equations you will find that: $a = 2$, $b = 3$, $c = -4$, and $d = -6$; leading us to:

$$\begin{bmatrix} -5 & 2 \\ 9 & -4 \end{bmatrix}\begin{bmatrix} 2 & 3 \\ -4 & -6 \end{bmatrix} = \begin{bmatrix} 1 & 0 \\ 0 & 1 \end{bmatrix}$$

Also, as you can easily check: $\begin{bmatrix} 2 & 3 \\ -4 & -6 \end{bmatrix}\begin{bmatrix} -5 & 2 \\ 9 & -4 \end{bmatrix} = \begin{bmatrix} 1 & 0 \\ 0 & 1 \end{bmatrix}$.

As for $\begin{bmatrix} 2 & 3 \\ -4 & -6 \end{bmatrix}$:

CHECK YOUR UNDERSTANDING 3.5

Verify that $\begin{bmatrix} 2 & 3 \\ -4 & -6 \end{bmatrix}$ is not a unit in $M_{2\times 2}$.

Answer: See page A-18.

THEOREM 3.2 Let R be a ring with unity 1. Then:

(a) $(-1)(-1) = 1$

(b) For any $a \in R$: $(-1)a = a(-1) = -a$

PROOF: (a) $(-1)(-1) \underset{\uparrow}{=} (1)(1) = 1$
              Theorem 3.1(c)

(b) $(-1)a \underset{\uparrow}{=} 1(-a) = -a$ and $a(-1) \underset{\uparrow}{=} -[a(1)] = -a$
      Theorem 3.1(b)   Theorem 3.1(b)

You are invited to establish the following result in the exercises:

THEOREM 3.3 For any integer $n > 1$, $\langle Z_n, +_n, \cdot_n \rangle$, under addition and multiplication modulo n; which is to say, for $a, b \in Z_n$:

$$a +_n b = r \text{ where: } a + b = qn + r, 0 \leq r < n$$
and (see margin)
$$a \cdot_n b = r \text{ where: } ab = qn + r, 0 \leq r < n$$

is a ring.

For example, in
$Z_6 = \{0, 1, 2, 3, 4, 5\}$:
$1 +_6 3 = 4$ and $2 +_6 5 = 1$
while
$1 \cdot_n 3 = 3$ and $2 \cdot_n 5 = 4$

CHECK YOUR UNDERSTANDING 3.6

(a) Determine the units in the ring Z_6

(b) Show that $m \in Z_n$ is a unit if and only if m and n are relatively prime.

Answer: (a) 1, 5
(b) See page A-18.

As might be anticipated:

DEFINITION 3.4 A **subring** of a ring R is a nonempty subset S of R which is itself a ring under the imposed binary operations of R.

As was the case with groups, the above subring definition can be recast in a more compact form:.

THEOREM 3.4 Let $\langle R, +, \cdot \rangle$ be a ring. A subset S of R is a subring of R if and only if:

(i) $\langle S, + \rangle$ is a subgroup of $\langle R, + \rangle$.

(ii) S is closed under multiplication, i.e:
$$s, \bar{s} \in S \Rightarrow s\bar{s} \in S$$

PROOF: If S is a subring of R then (i) and (ii) clearly hold. Conversely, if (i) and (ii) hold then, since $a(bc) = (ab)c$ along with $a(b + c) = ab + ac$ and $(a + b)c = ac + bc$ hold for all elements $a, b, c \in R$, they must surely hold for all elements $a, b, c \in S$.

CHECK YOUR UNDERSTANDING 3.7

Let $\langle R, +, \cdot \rangle$ be a ring. A subset S of R is a subring of R if and only if for every $s, \bar{s} \in S$:

(i) $s - \bar{s} \in S$ and (ii) $s\bar{s} \in S$

Suggestion: Consider Exercise 38, page 70

Answer: See page A-18

EXAMPLE 3.4 (a) Show that, for any $n \in Z^+$, the additive group nZ under standard multiplication is a subring of Z.

(b) Is $U_{2 \times 2} = \{A | A \text{ is a unit in } M_{2 \times 2}\}$ a subring of $M_{2 \times 2}$?

(a) For any $a, b \in Z$:
$$na - nb = n(a-b) \in nZ, \text{ and } (na)(nb) = n(nab) \in nZ.$$

(b) No. $U_{2 \times 2}$ is not closed under addition:

Incidentally $\langle U_{2 \times 2}, \cdot \rangle$ is a group (under multiplication) (Exercise 40)

$$\begin{bmatrix} 1 & 0 \\ 0 & 1 \end{bmatrix}, \begin{bmatrix} -1 & 0 \\ 0 & -1 \end{bmatrix} \in U_{2 \times 2} \text{ but } \begin{bmatrix} 1 & 0 \\ 0 & 1 \end{bmatrix} + \begin{bmatrix} -1 & 0 \\ 0 & -1 \end{bmatrix} = \begin{bmatrix} 0 & 0 \\ 0 & 0 \end{bmatrix} \notin U_{2 \times 2}.$$

CHECK YOUR UNDERSTANDING 3.8

(a) Show that $H = \left\{ \begin{bmatrix} 0 & 0 \\ a & b \end{bmatrix} \middle| a, b \in \Re \right\}$ is a subring of $M_{2 \times 2}$.

(b) Let a be and element of a ring R. Show that
$$S_a = \{x \in R | ax = 0\}$$
is a subring of R.

Answer: See page A-18.

118   Part 3   From Rings To Fields

	EXERCISES	

Exercise 1-12. Determine if the given set it a ring under the give addition and multiplication operations. If it is a ring, indicated whether or not it is commutative, and whether or not it has a unity.

1. The set nZ under standard addition and multiplication.

2. The set $\{2n | n \in Z^+\}$ of positive even integers under standard addition and multiplication.

3. The set $\{2n | (n \geq 0)\}$ of nonnegative even integers under standard addition and multiplication.

4. The set $a + b\sqrt{2} | (a, b \in \Re)$ under standard addition and multiplication.

5. The set $a + b\sqrt{2} | (a, b \in Q)$ under standard addition and multiplication.

6. The set $\{0, 1\}$ under standard addition and multiplication.

7. The set $2Z \times Q$ under component addition and multiplication.

8. The set $2Z \times \{0, 1\}$ under component standard addition and multiplication.

9. The set $\left\{ \begin{bmatrix} a & b \\ 0 & 0 \end{bmatrix} \middle| a, b, c \in \Re \right\}$ under matrix addition and multiplication. (See Example 3.2.)

10. The set $\left\{ \begin{bmatrix} a & b \\ c & 0 \end{bmatrix} \middle| a, b, c \in \Re \right\}$ under matrix addition and multiplication. (See Example 3.2.)

11. The set $\left\{ \begin{bmatrix} a & b \\ 0 & c \end{bmatrix} \middle| a, b, c \in \Re \right\}$ under matrix addition and multiplication. (See Example 3.2)

12. The set of polynomials, $p(x)$, with real coefficients, of degree less than or equal to 5, under standard polynomial addition and multiplications.

Exercise 13-20. Determine if the given subset S of the giver ring R is a subring of R.

13. $R = Q$, and $S = Q^+$.

14. $R = Q$ and $S = Z$.

15. $R = Z \times Z$ and $S = \{(n, n)\}$.

16. $R = Q$, and $S = \{q^2 | q \in Q\}$.

17. $R = Z \times Z$ and $S = \{(n, n) | n \leq 0\}$.

18. $R = Z \times Z$ and $S = \{(n, 2n)\}$.

19. $R = M_{2 \times 2}$ and $S = \left\{ \begin{bmatrix} a & a \\ 0 & 0 \end{bmatrix} \right\}$.

20. $R = M_{2 \times 2}$ and $\begin{bmatrix} a & a-b \\ a-b & b \end{bmatrix}$.

Exercise 21-27. Find the units in the give ring.

21. Z 22. $5Z$ 23. Z_5 24. Z_{15} 25. $Z \times Z$ 26. $Z \times Q$ 27. $Z_6 \times Z_9$

28. Show that any abelian group $\langle G, + \rangle$ can be turned into a ring by defining $ab = 0$ for every $a, b \in G$.

29. Verify that for any $A, B, C \in M_{2 \times 2}$ (see Example 3.2):

    (a) $(AB)C = A(BC)$ (b) $A(B + C) = AB + AC$ (c) $(A + B)C = AC + BC$

30. Let R_1 and R_2 be rings. Prove that $R_1 \cap R_2$ is a ring.

31. Let $\{R_i\}_{i=1}^n$ be a collection of rings. Prove that $\bigcap_{i=1}^n R_i$ is a ring.

32. Let $\{R_\alpha\}_{\alpha \in A}$ be a collection of rings. Prove that $\bigcap_{\alpha \in A} R_\alpha$ is a ring.

33. Let a and b be element in a ring R. Show that $nm(ab) = (na)(mb)$ for any integer n and m.

34. Describe all of the subrings of the ring of integer.

35. Let the ring R be cyclic under addition. Prove that R is commutative.

36. Let $F(\Re)$ denote the set of all real-valued functions. For f and g in $F(\Re)$, let $f + g$ be given by $(f + g)(x) = f(x) + g(x)$ and $(fg)(x) = f(x)g(x)$. Show that under these operation $F(\Re)$ is a ring with unity.

37. The **center** of a ring R is the set $\{x \in R | ax = xa \; \forall a \in R\}$. Sow that the center of R is a subring of R.

38. For a and element of a ring R, let $C(a) = \{x \in R | xa = ax\}$. Show that $C(a)$ is a subring of R containing a.

39. Show that the center of a ring R is equal to $\bigcap_{a \in R} C(a)$. (See Exercises 36 and 37.)

40. Prove that $\begin{bmatrix} a & b \\ c & d \end{bmatrix}$ is a unit in $M_{2 \times 2}$ if and only if $ad - bc \neq 0$.

41. Prove that if $a \in R$ is a unit, then it has a unique inverse.

42. Prove that the set $U_{2 \times 2} = \{A | A \text{ is a unit in } M_{2 \times 2}\}$ is a group under multiplication.

43. Let R be a ring, and let $a \in R$. Show that the set $S_a = \{axa | x \in R\}$ is a subring of R.

44. Show that the multiplicative inveres of any unit in a ring with unity is unique.

45. Let R be a commutative ring with unity, and let $U(R)$ denote the set of units in R. Prove that $U(R)$ is a group under the multiplication of R.

46. Show that if there exists an integer n greater than 1 for which $x^n = x$ for every element x in a ring R, then $ab = 0 \Rightarrow ba = 0$.

47. Let k be the least common multiple of the positive integers m and n. Show that $m\mathbb{Z} \cap n\mathbb{Z} = k\mathbb{Z}$.

48. Let R be a commutative ring. Prove that $a^2 - b^2 = (a+b)(a-b)$.

49. An element a of a ring R is **idempotent** if $a^2 = a$. Show that the set of all idempotent elements of a commutative ring is closed under multiplication.

50. An element a of a ring R is **nilpotent** if $a^n = 0$ for some $n \in \mathbb{Z}^+$. Show that if a and b are nilpotent elements of a commutative ring R, then $a + b$ is also nilpotent.

51. A ring R is said to be a **Boolean ring** if $a^2 = a$ for every $a \in R$. Prove that every Boolean ring is commutative.

52. Give an example of finite Boolean ring, and an example of an infinite Boolean ring (see Exercise 50).

53. Prove Theorem 3.3.

54. Prove that m is a unit in Z_n if and only if $gcd(n, m) = 1$.

55. Let $R_1, R_2, \ldots, R_n$ be rings. Show that:

    (a) $R_1 \times R_2 \times \ldots \times R_n$ with operations
    $$(a_1, a_2, \ldots, a_n) + (b_1, b_2, \ldots, b_n) = (a_1 + b_1, a_2 + b_2, \ldots, a_n + b_n)$$
    $$(a_1, a_2, \ldots, a_n)(b_1, b_2, \ldots, b_n) = (a_1 b_1, a_2 b_2, \ldots, a_n b_n)$$
    is a ring.

    (b) $R_1 \times R_2 \times \ldots \times R_n$ is commutative if and only if R_i is commutative for $1 \le i \le n$.

    (c) $R_1 \times R_2 \times \ldots \times R_n$ has a unity if and only if R_i has a unity for $1 \le i \le n$.

Prove or Give a Counterexample

56. If R_1 and R_2 are rings, then $R_1 \cup R_2$ is a ring.

57. In any ring R, $ab = 0 \Rightarrow ba = 0$.

58. If $x^3 = x$ for all elements x in a ring R, then $6x = 0$ for all $x \in R$.

59. In any ring R: $a^2 - b^2 = (a+b)(a-b)$.

60. If R_1 and R_2 are Boolean ring, then $R_1 \cap R_2$ is a Boolean ring. (See Exercise 50).

61. If R_1 and R_2 are Boolean ring, then $R_1 \times R_2$ is a Boolean ring. (See Exercise 50).

62. The set of all idempotent elements in a ring R is a subring of R. (See Exercise 48).

§2. HOMOMORPHISMS AND QUOTIENT RINGS

Moving the group-homomorphism concept of page 72 up a notch we come to::

> So, a ring homomorphism preserves both the sum and product operations:
> You can perform sums and products in R and then carry the results over to the ring R' (via ϕ), or you can first carry a and b over to R' and then perform the operations in that ring.

DEFINITION 3.5
RING HOMOMORPHISM

The function $\phi: \langle R, +, \cdot \rangle \to \langle R', +, \cdot \rangle$ is a **homomorphism** if, for every $a, b \in R$:

(1) $\phi(a+b) = \phi(a) + \phi(b)$

and (2) $\phi(ab) = \phi(a)\phi(b)$

ISOMORPHISM

A homomorphism $\phi: R \to R'$ which is also a bijection is said to be an **isomorphism** from the ring R to the ring R'.

ISOMORPHIC

Two rings R and R' are **isomorphic**, written $R \cong R'$, if there exists an isomorphism from one of the rings to the other.

Condition (1) above assures us that a ring homomorphism is also a group homomorphism $\phi: \langle R, + \rangle \to \langle R', + \rangle$. That being the case, previously encountered group-homomorphic results remain in effect in the current setting. In particular, Theorem 2.25, page 76, tells us that:

> Why $\text{Ker}(\phi) = \{0\}$ rather than $\text{Ker}(\phi) = \{e\}$?
> Because we are dealing with abelian groups $\langle R, + \rangle$ and $\langle R', + \rangle$ that's why.

> A ring homomorphism $\phi: \langle R, +, \cdot \rangle \to \langle R', +, \cdot \rangle$ is one-to-one if and only if $\text{Ker}(\phi) = \{0\}$.

EXAMPLE 3.5 Let $\phi: Z \to Z_n$ be given by $\phi(a) = r_a$, where $a = q_a n + r_a$, with $0 \leq r_a < n$. Show that ϕ is a ring homomorphism.

SOLUTION: A consequence of CYU 1.18, page 36, and the fact that:

$$\phi(a+b) \equiv [\phi(a) + \phi(b)] \bmod n \text{ and } \phi(ab) \equiv \phi(a)\phi(b) \bmod n$$

As it is with group homomorphisms, we have:

THEOREM 3.5 Let $\phi: R \to R'$ be a ring homomorphism.

(a) If H is a subring of R, then $\phi(H)$ is a subring of R'.

(b) If H' is a subring of R', then $\phi^{-1}(H')$ is a subring of R.

PROOF: We establish (a) and invite you to verify (b) in CYU 3.9.

Appealing to CYU 3.7, page 116, we show that the nonempty set $\phi(H)$ is closed under subtraction and multiplication:

$$\phi(h_1) - \phi(h_2) = \phi(h_1 - h_2) \in \phi(H)$$

$$\phi(h_1)\phi(h_2) = \phi(h_1 h_2) \in \phi(H)$$

CHECK YOUR UNDERSTANDING 3.9

(a) Let $\phi: R \to R'$ be a homomorphism. Prove that if H' is a subring of R', then $\phi^{-1}(H')$ is a subring of R.

(b) Show that the rings $3Z$ and $5Z$ are not isomorphic.
(Compare with CYU 2.22(b), page 77)

Answer: See page A-19.

Every ring $\langle R, +, \cdot \rangle$ is, in part, an abelian group: $\langle R, + \rangle$. Choosing to use the sum notation in Theorem 2.39 (page 95) we have:

> For any subring H of the ring $\langle R, +, \cdot \rangle$, the factor group $G/H = \{a+H\}_{a \in G}$ is a group under the coset operation:
> $(a+H) + (b+H) = (a+b) + H$

Fine, but will that factor group G/N evolve into a ring under the "natural" product operation $(a+H)(b+H) = (ab) + H$? Not necessarily. Indeed, that coset "product" need not be defined. A case in point:

EXAMPLE 3.6

Consider the subring $H = \left\{ \begin{bmatrix} 0 & 0 \\ a & b \end{bmatrix} \middle| a, b \in \mathfrak{R} \right\}$ of CYU 3.8(a), page 117. Show that

$$\left(\begin{bmatrix} 1 & 0 \\ a & b \end{bmatrix} + H \right) \left(\begin{bmatrix} 0 & 1 \\ a & b \end{bmatrix} + H \right) = \left(\begin{bmatrix} 1 & 0 \\ a & b \end{bmatrix} \begin{bmatrix} 0 & 1 \\ a & b \end{bmatrix} \right) + H$$

is not a well defined operation.

SOLUTION: To be well define, the set products must yield the same result, independently of the chosen representative for the two given cosets. However, while:

$\begin{bmatrix} 1 & 0 \\ 0 & 0 \end{bmatrix} - \begin{bmatrix} 1 & 0 \\ 0 & 1 \end{bmatrix} = \begin{bmatrix} 0 & 0 \\ 0 & 1 \end{bmatrix} \in H$

$\begin{bmatrix} 1 & 0 \\ 0 & 0 \end{bmatrix} + H \underset{\text{margin}}{\overset{\uparrow}{=}} \begin{bmatrix} 1 & 0 \\ 0 & 1 \end{bmatrix} + H$ and $\begin{bmatrix} 0 & 1 \\ 0 & 0 \end{bmatrix} + H = \begin{bmatrix} 0 & 1 \\ 0 & 1 \end{bmatrix} + H$:

$\begin{bmatrix} 1 & 0 \\ 0 & 0 \end{bmatrix} \begin{bmatrix} 0 & 1 \\ 0 & 0 \end{bmatrix} + H$ is **not** equal to $\begin{bmatrix} 1 & 0 \\ 0 & 1 \end{bmatrix} \begin{bmatrix} 0 & 1 \\ 0 & 1 \end{bmatrix} + H$.

Why not? Because: $\begin{bmatrix} 1 & 0 \\ 0 & 0 \end{bmatrix} \begin{bmatrix} 0 & 1 \\ 0 & 0 \end{bmatrix} = \begin{bmatrix} \mathbf{0} & \mathbf{1} \\ \mathbf{0} & \mathbf{0} \end{bmatrix}$, $\begin{bmatrix} 0 & 1 \\ 0 & 0 \end{bmatrix} \begin{bmatrix} 0 & 1 \\ 0 & 1 \end{bmatrix} = \begin{bmatrix} \mathbf{0} & \mathbf{2} \\ \mathbf{0} & \mathbf{0} \end{bmatrix}$

and $\begin{bmatrix} \mathbf{0} & \mathbf{1} \\ \mathbf{0} & \mathbf{0} \end{bmatrix} - \begin{bmatrix} \mathbf{0} & \mathbf{2} \\ \mathbf{0} & \mathbf{0} \end{bmatrix} = \begin{bmatrix} 0 & -1 \\ 0 & 0 \end{bmatrix} \notin H$

Note that the factor **group** R/H exists, as every subgroup of an abelian group is normal.

The above example illustrates the fact that for a given subring H of a ring R, one can not expect that the factor group R/H of Theorem 2.39, page 95, will become a ring under the (attempted) product $(a+H)(b+H) = ab+H$. Of particular importance are those subring for which that expectation will be realized:

DEFINITION 3.6 A subring I or a ring R is a (two-sided) **ideal**
IDEAL if for any $a \in I$ and every $r \in R$:
$$ra \in I \text{ and } ar \in I.$$

Justifying our expectation:

THEOREM 3.6 If I is an ideal in R then the (additive) factor group R/I turns into a ring under the imposed multiplication $(a+I)(b+I) = (ab)+I$.

R/I is said to be the **quotient ring of R by N**. Quotient rings are also said to be factor rings.

PROOF: The first order of business is to show that the above coset product operation is well defined; which is to say that:

If $a+I = a'+I$ and $b+I = b'+I$ then: $ab+I = a'b'+I$.

Lets do it:

$$a+I = a'+I \Rightarrow a-a' \in I \underset{\substack{\uparrow \\ \text{since } I \text{ is an ideal}}}{\Rightarrow} (a-a')b \in I \Rightarrow ab - a'b \in I$$

$$b+I = b'+I \Rightarrow b-b' \in I \overset{\downarrow}{\Rightarrow} a'(b-b') \in I \Rightarrow a'b - a'b' \in I$$

Since both $ab - a'b$ and $a'b - a'b'$ are in I:

$$(ab - a'b) + (a'b - a'b') = ab - a'b' \in I$$

Thus $ab + I = a'b' + I$

(as I, being a subgroup of an abelian group, is normal.

As for the ring part of the proof, we need only establish the associative and distributive axioms of Definition 3.1 (page 111), as we already know that R/I is an abelian group. Not a serious challenge, now that we know that the coset multiplications is well defined:

Associative: $(a+I)[(b+I)(c+I)] = (a+I)[(bc)+I]$
$$= a(bc) + I = (ab)c + I$$
$$= [(a+I)(b+I)](c+I)$$

Distributive:
$$(a+I)[(b+I) + (c+I)] = (a+I)[(b+c)I]$$
$$= [a(b+c)] + I$$
$$= (ab + ac) + I$$
$$= (ab + I) + (ac + I)$$
$$= (a+I)(b+I) + (a+I)(c+I)$$

In the same fashion, one can show that:
$$[(b+I) + (c+I)](a+I) = (b+I)(a+I) + (c+I)(a+I)$$

CHECK YOUR UNDERSTANDING 3.10

(a) Show that I is an ideal in Z if and only if $I = nZ$.

(b) Let $\phi: R \to R'$ be an onto ring homomorphism. Show that if I is an ideal in R then $\phi(I)$ is an ideal in R'

Answer: See page A-19.

Roughly speaking:

> NORMAL SUBGROUPS ARE TO GROUPS
> AS
> IDEALS ARE TO RINGS

A case in point (compare with Theorem 2.43, page 98):

THEOREM 3.7
FIRST ISOMORPHISM THEOREM

If $\phi: R \to R'$ is a ring homomorphism, then $K = \text{Ker}(\phi)$ is an ideal of R and:
$$R/K \cong \phi(R)$$

PROOF: We already know that $K = \text{Ker}(\phi) = \phi^{-1}\{0\}$ is an additive subgroup of R. It is, in fact, an ideal since, for any $a \in K$ and every $r \in R$, both ra and ar are in K:

$$\phi(ra) = \phi(r)\phi(a) = \phi(r)(0) = 0 \text{ and } \phi(ar) = 0$$

The proof of Theorem 2.43 serves to show that the function
$$\psi: R/K \to \phi(R) \text{ given by } \psi(r + K) = \phi(r)$$
is a group isomorphism. Indeed, it a ring isomorphism:

$$\psi[(r_1 + K)(r_2 + K)] = \psi[(r_1 r_2) + K]$$
$$= \phi(r_1 r_2) = \phi(r_1)\phi(r_2)$$
$$= \psi[(r_1 + K)]\psi[(r_2 + K)]$$

CHECK YOUR UNDERSTANDING 3.11

Let $\phi: R \to R'$ be an onto ring homomorphism with kernel K. If I' is an ideal of R', then $I = \varphi^{-1}(I')$ is an ideal in R containing K and:
$$I/K \cong I'$$

Answer: See page A-19.

In CYU 3.10(a) you were invited to show that, under standard addition and multiplication, nZ is an ideal of Z. More can be said:

EXAMPLE 3.7 Show that for any positive integer n:
$$Z_n \cong Z/(nZ)$$

SOLUTION: In Example 2.6, page 72, we showed that the function $\phi: Z \to Z_n$ given by $\phi(m) = r$ where $m = nq + r$ with $0 \leq r < n$ is a group homomorphism. You are invited to show, in CYU 3.12, that the function also preservers products; in other words, that it is a ring homomorphism from the ring Z to the ring Z_n.

While ϕ is not one-to-one, it is certainly onto. As such we know, by Theorem 3.7, that:
$$Z_n \cong Z/K \text{ where } K = \text{Ker}(\phi).$$
Noting that:
$$\text{Ker}(\phi) = \{m|\phi(m) = 0\} = \{kn|k \in Z\} = nZ$$
↑
Example 2.4, page 62

we conclude that: $Z_n \cong Z/(nZ)$.

CHECK YOUR UNDERSTANDING 3.12

Let $\phi: Z \to Z_n$ be given by $\phi(m) = r$ where $m = nq + r$ with $0 \le r < n$. Show that for any $a, b \in Z$: $\phi(ab) = \phi(a)\phi(b)$.

Answer: See page A-19.

You are invited to establish the next two isomorphism theorems in the exercises.

THEOREM 3.8
SECOND ISOMORPHISM THEOREM

Let H be a subring of a ring R, and I an ideal of R. Then:
$$H + I = \{h + i | h \in H, i \in I\}$$
is a subring of R, I is an ideal of $H + I$, and:
$$(H + I)/I \cong H/(H \cap I)$$

(Compare with Theorem 2.44, page 99.)

THEOREM 3.9
THIRD ISOMORPHISM THEOREM

Let $\phi: R \to R'$ be an onto homomorphism with kernel K. If I' is an ideal of R', then:
$$I = \phi^{-1}(I') = \{a \in R | \phi(a) \in I'\}$$
is ideal of R and:
$$R/I \cong R'/(I')$$

(Compare with Theorem 2.45, page 99.)

EXERCISES

Exercise 1-6. Determine if the given map $\varphi: R \to R'$ is a ring homomorphism.

1. $R = R' = Z$, and $\phi(n) = 3n$.
2. $R = Z$, $R' = 3Z$ and $\phi(n) = 3n$.
3. $R = R' = M_{2 \times 2}$ and $\phi \begin{bmatrix} a & b \\ c & d \end{bmatrix} = \begin{bmatrix} -d & b \\ d & c \end{bmatrix}$.
4. $R = R' = M_{2 \times 2}$ and $\phi \begin{bmatrix} a & b \\ c & d \end{bmatrix} = \begin{bmatrix} ab & 0 \\ 0 & cd \end{bmatrix}$.
5. $R = M_{2 \times 2}$, $R' = \Re$ and $\phi \begin{bmatrix} a & b \\ c & d \end{bmatrix} = a$.
6. $R = M_{2 \times 2}$, $R = \Re$ and $\phi \begin{bmatrix} a & b \\ c & d \end{bmatrix} = ad - bc$.

Exercise 7-10. Determine if the given subset S of the ring R is an ideal of R.

7. $R = M_{2 \times 2}$, $S = \left\{ \begin{bmatrix} a & 0 \\ 0 & d \end{bmatrix} \middle| a, d \in \Re \right\}$.
8. $R = M_{2 \times 2}$, $S = \left\{ \begin{bmatrix} a & 1 \\ 0 & d \end{bmatrix} \middle| a, d \in \Re \right\}$.
9. $R = Z \times Z$, $S = \{(a, -a) | a \in Z\}$.
10. $R = Z \times Z$, $S = \{(2a, a) | a \in Z\}$.

11. Let $\phi: R \to R'$ be a homomorphism from R onto R'. Show that:
    (a) $\phi(a^n) = [\phi(a)]^n$ for all $a \in R$ and $n > 0$.
    (b) If R possesses a unity then so does the ring $\phi(R)$.
    (c) If a is a unit of R, then $\phi(a)$ is a unit of $\phi(R)$.

12. Let $\phi: R \to R'$ and $\theta: R' \to R''$ be ring homomorphisms. Prove that the composite function $\theta \circ \phi: R \to R''$ is also a homeomorphism.

13. Let S be a subset of a ring R. Show that S is an ideal of R if and only if the following two conditions hold:
    (i) S is an additive subgroup of R.
    (ii) For every $s \in S$ and $r \in R$ we have $rs \in S$ and $sr \in S$.

14. Let $F(\Re)$ denote the ring of all real-valued functions of Exercise 36, page 119, and let $a \in \Re$. Show that the map $\phi_a: F(\Re) \to \Re$ given by $\phi_a(f) = f(a)$ is a homomorphism.

15. Let I be an ideal is commutative ring R with unity 1. Show that R/I is a commutative ring with unity.

16. Let R be a ring with unity. Show that if I is an idea of R that contains a unit, then $I = R$.

17. Let I be an ideal in a ring R. Show that there exist an onto ring homomorphism $\phi: R \to R/I$ with $\text{Ker}(\phi) = I$.

18. Let I be an ideal in a ring R. Show that if K is an ideal in I, then $\overline{K} = \{a + I | a \in K\}$ is an ideal in R/I.

19. Let I be an ideal in a ring R. Show that if $\bar{K}$ is an ideal in R/I, then there exists an ideal K in R with $I \subseteq K$ such that $\bar{K} = K/I = \{a + I | a \in K\}$.

20. Describe all ring homomorphisms from Z to Z.

21. Describe all ring homomorphisms from Z to $Z \times Z$.

22. Show that if $n \neq m$, then the rings nZ and mZ are not isomorphic.

23. Prove that $\cong$ (isomorphic) is an equivalence relation on any set of rings (see Definition 1.12, page 29).

24. Show that the function $\phi: Z \to Z_n$ given by $\phi(m) = r$ where $m = nq + r$ with $0 \leq r < n$ preserves products: $\phi(st) = \phi(s)\phi(t) \; \forall s, t \in Z$.

25. Find a subring of $Z \times Z$ that is not an ideal of $Z \times Z$.

26. Prove that I is an ideal of a ring R if and only if:
    (i) $0 \in I$
    (ii) If $a, b \in I$, then $a - b \in I$
    (iii) If $a \in I$ and $r \in R$, then $ra \in R$

27. An element a of a ring R is **nilpotent** if $a^n = 0$ for some $n \in Z^+$. Show the collection of nilpotent elements of commutative ring R is an ideal of R.

28. Let R be a commutative ring and $a \in R$. Show that $\{x \in R | xa = 0\}$ is an ideal of R.

29. Prove that if I_1 and I_2 are ideals of R, then $I_1 \cap I_2$ is an ideal of R.

30. Let A and B be ideals of R, and let I be the set of all elements of the form ab with $a \in A$ and $b \in B$. Prove that I is an ideal of R.

31. Let H be a subring or R that is not an ideal of R. Verify that the operation $(a + H)(b + H) = ab + H$ is not well defined..

32. Prove Theorem 3.8.

33. Prove Theorem 3.9.

PROVE OR GIVE A COUNTEREXAMPLE

34. If I_1 and I_2 are ideals of R, then $I_1 \cup I_2$ is an ideal of R.

35. If I is an ideal of R and if H is a subring of R, then $I \cap H$ is an ideal of R.

36. For every element a of a ring R, the set $\{x \in R | xa = 0\}$ is an ideal of R.

37. Let $\phi: R \to R'$ be a homomorphism from R onto R'. Show that if R possesses a unity then so does R'.

38. The collection of nilpotent elements n a ring R is an ideal of R. (See Exercise 27).

§3. INTEGRAL DOMAINS AND FIELDS

The set Z of integers under addition led to the definition of a group on page 41. Tossing multiplication into the mix brought us to the definition of a ring on page 111. How about "division"? Can one perform (grade school) division in the ring Z, or Q, or $\Re$? Absolutely not in Z, where you can only divide by 1 or -1. Q and $\Re$ fair much better in that one can divide by any number other then 0. As it turns out, Q and $\Re$ are examples of fields:

DEFINITION 3.7
ZERO DIVISOR — A **zero-divisor** in a commutative ring R is a nonzero element a for which there exits a nonzero element $b \in R$ with $ab = 0$.

INTEGRAL DOMAIN — An **integral domain** is a commutative ring R with unity that contains no zero-divisors.

Note that Every field is an integral domain.

FIELD — A **field** is a commutative ring with unity in which every nonzero element is a unit.

As is depicted below, fields are at the top of our algebraic pecking order, and groups are at the bottom:

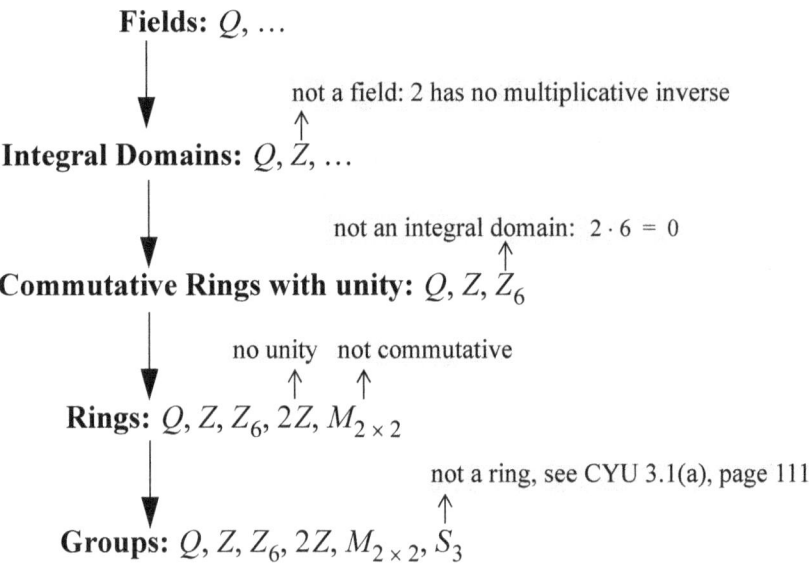

CHECK YOUR UNDERSTANDING 3.13

Assign to the given group its highest algebraic rank (Field at the top, and Group at the bottom).

(a) Z_5 (b) Z_{15}

Answer: (a) Field. (b) Commutative ring with unity.

3.3 Integral Domains and Fields

The familiar high school cancellation law (margin) holds in any integral domain:

$\dfrac{ab}{ac} = \dfrac{b}{c}$ (if $a \neq 0$)

THEOREM 3.10 Let D be an integral domain, and $a, b, c \in D$.

If $ab = ac$ and $a \neq 0$, then $b = c$

PROOF: $ab = ac \Rightarrow ab - ac = 0 \Rightarrow a(b-c) = 0$. Since D is an integral domain, and since $a \neq 0$: $b - c = 0$.

CHECK YOUR UNDERSTANDING 3.14

(a) Prove that a commutative ring with unity is an integral domain if and only if the cancellation property of Theorem 3.10 holds.

(b) Let D be an integral domain, and let $a \neq 0$ be an element in D. Show that the function $f_a: D \to D$ given by $f_a(x) = ax$ is one-to-one.

Answer: See page A-20.

THEOREM 3.11 A commutative ring with unity R is a field if and only if $\{0\}$ and R are the only ideals in R.

PROOF: Let R be a field and let I be an ideal in R with $I \neq \{0\}$. Chose $a \in I$, $a \neq 0$. Since R is a field and I is an ideal, we then have: $a^{-1}a = 1 \in I$. Since I is an ideal: $\{r \cdot 1 | r \in R\} = R \subseteq I$. Thus: $I = R$.

To establish the converse, we show that if $\{0\}$ and R are the only ideals in R, then every nonzero element in R is a unit:

Let $a \neq 0$ be an element of R. Consider the ideal $I = \langle a \rangle$. Since $I \neq \{0\}$, $I = R$. We then have $1 \in I = \langle a \rangle$. It follows that $a^n = 1$ for some $n \in Z$, and that a^{n-1} is the inverse if a.

THEOREM 3.12 Any finite Integral domain is a field.

PROOF: Let D be a finite integral domain, and let $a \neq 0$ be an element in D. CYU 3.14(b) assures us that the function $f_a: D \to D$ given by $f_a(x) = ax$ is one-to-one. It follows, since D is finite, that the function is also onto. In particular, there must exist some $b \in D$ such that $ab = 1$, and a is seen to be a unit. Since a was an arbitrary nonzero element in D, D is a field.

CHECK YOUR UNDERSTANDING 3.15

Prove that for every prime p, Z_p is a field.

Answer: See page A-20.

We focus briefly, and somewhat loosely, on the set $P_Z[x]$ of polynomials with integer coefficients, as well as the sets $P_{Z_n}[x]$ of polynomials with coefficients taken from the rings $Z_n = \{0, 1, 2, ..., n-1\}$. All turn out to be commutative rings (with unity) under the following standard sum and product operations:

In $P_Z[x]$	In $P_{Z_6}[x]$
$(3x^2 + 4x + 5) + (x^2 + 4x + 1) = 4x^2 + 8x + 6$	$(3x^2 + 4x + 5) + (x^2 + 4x + 1) = 4x^2 + 2x$
$(2x^2 - 4x)(5x - 2) = 2x^2(5x - 2) - 4x(5x - 2)$ $= 10x^3 - 4x^2 - 20x^2 + 8x$ $= 10x^3 - 24x^2 + 8x$	$(2x^2 - 4x)(5x - 2) = 2x^2(5x - 2) - 4x(5x - 2)$ $= 4x^3 - 4x^2 - 2x^2 + 2x$ $= 4x^3 + 2x$

Note that while the coefficients of a polynomial $p(x)$ in $P_{Z_n}[x]$ are elements of the ring $Z_n = \{0, 1, 2, ..., n-1\}$, the degree of such a polynomial can be any nonnegative integer. In particular, while $3x^9 - 5x^2 + x - 1$ might very well be a polynomial in $P_{Z_6}[x]$ (of degree 9), it can not be a polynomial in $P_{Z_5}[x]$ ($5 \notin Z_5$).

Consider the polynomial $x^2 - x - 6$. Since the distributive property holds in both $P_Z[x]$ and in $P_{Z_{12}}[x]$ we can express the polynomial in factored form in either ring:

$$x^2 - x - 6 = (x - 3)(x + 2)$$

But while the equation:

$$x^2 - x - 6 = (x - 3)(x + 2) = 0$$

has but two solutions in $P_Z[x]$ (3 and -2), the same equation turns out to have four solutions in $P_{Z_{12}}[x]$. The reason, you see, is that while the ring Z is an integral domain (no zero divisors), the same cannot be said for Z_{12}. Indeed there are several pairs on nonzero elements in Z_{12} with product equal to zero:

$$2 \cdot 6 = 0, \ 3 \cdot 4 = 0, \ 8 \cdot 3 = 0, \ 9 \cdot 4 = 0, \ 10 \cdot 6 = 0$$
$$24 \equiv 0 \bmod 12$$

Your turn:

CHECK YOUR UNDERSTANDING 3.16

Solve the equation $x^2 - x - 6 = 0$ in:

(a) Z_{12} (b) Z_8

Answer: (a) 3, 6, 7, 10
(b) 3, 6

3.3 Integral Domains and Fields 131

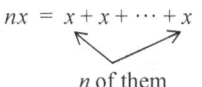

DEFINITION 3.8
CHARACTERISTIC

The **characteristic** of a ring R is the least positive integer n such that $nx = 0$ for every $x \in R$. If no such integer exists, then R is said to have characteristic 0.

The ring Z has characteristic 0, and the cyclic ring Z_n has characteristic n. Clearly no finite ring is of characteristic 0. Must every infinite ring have characteristic 0? No:

CHECK YOUR UNDERSTANDING 3.17

Answer: See page A-21.

Prove that the infinite ring $P_{Z_n}[x]$ has characteristic n.

To determine the characteristic of a ring with unity, one need look no further than its unity:

THEOREM 3.13 Let R be a ring with unity. If $n1 \neq 0$ for all $n \in Z^+$, then R has characteristic 0. If $n1 = 0$ for some $n \in Z^+$, then the smallest such n is the characteristic of R.

PROOF: If $n1 \neq 0$ for all $n \in Z^+$ then surely there cannot exist $n \in Z^+$ such that $nx = 0$ for every $x \in R$, and R has characteristic 0. On the other hand, if $n1 = 0$ for some positive integer n, then, for any $x \in R$:

$$nx = nx = \underbrace{x + x + \cdots + x}_{n \text{ of them}} = (1 + 1 + \ldots + 1)x = (n1)x = 0x = 0$$

The smallest such n is then the characteristic of R.

THEOREM 3.14 The characteristic of an integral domain D is either 0 or prime.

PROOF: Assume that D has positive characteristic n, and that n is not prime. Then n can be written as $n = st$ with $1 < s < n$ and $1 < t < n$. We then have:

$$0 = n1 = (st)1 = (st)1^2 = (s1)(t1)$$

Since D has no zero divisors, either $s1 = 0$ or $t1 = 0$. But this cannot occur, since n is the least positive integer such that $n1 = 0$. Conclusion: n must be prime.

CHECK YOUR UNDERSTANDING 3.18

Let D be an integral domain of characteristic 3. Show that for every $a, b \in D: (a+b)^3 = a^3 + b^3$.

In the exercises you invited to show that for any prime p, if D has characteristic p then:
$$(a+b)^p = a^p + p.$$

Answer: See page A-21.

PRIME AND MAXIMAL IDEALS

DEFINITION 3.9 Let R be a commutative ring.

PRIME IDEAL A **prime ideal** of R is a proper ideal I of R for which:
$$ab \in I \Rightarrow a \in I \text{ or } b \in I$$

Note: A **proper ideal** of R is, by definition, an ideal in R that is distinct for R itself.

MAXIMAL IDEAL A proper ideal I of R is a **maximal ideal** if R is the only ideal containing I.

EXAMPLE 3.8 (a) Show that an ideal I in Z is prime if and only if $I = pZ$, where p is a prime.

(b) Show that $5Z$ is a maximal ideal in Z.

SOLUTION: (a) Let p be prime. If $ab \in pZ$ then, by Theorem 1.9, page 24, $p|a$ or $p|b$; which is to say, that $a \in pZ$ or $b \in pZ$.

Conversely, assume that $I = nZ$ where $n > 1$ is not prime. Let $n = ab$ for some positive integers a and b. Then $ab \in I$ with neither a nor b in I (neither is a multiple of n).

(b) If I is an ideal properly containing $5Z$, then there must exists $a \in I$ with $a \notin 5Z$, i.e. 5 does not divide a. It follows, since 5 is prime, that $\gcd(5, a) = 1$. Employing Theorem 1.7, page 23, we have:
$$1 = 5s + at$$
for integers s and t. Since $5s$ and at are both in I: $1 \in I$. It follows, since I is an ideal in Z, that $I = Z$.

CHECK YOUR UNDERSTANDING 3.19

(a) Show that pZ is a maximal ideal for any prime p.

Answer: See page A-21.

(b) Prove that I is a maximal ideal in Z if and only if it is prime.

THEOREM 3.15 Let I be an ideal in a commutative ring R with unity.

The converse of both (a) and (b) also hold. See Exercise 25 and Exercise 26.

(a) If I is a prime ideal then R/I is an integral domain.

(b) If I is a maximal ideal then R/I is a field.

PROOF: (a) We need to show that R/I has no zero divisors; which is to say that if $ab + I = I$, then either $a + I = I$ or $b + I = I$; which is to say that if $ab \in I$ then either $a \in I$ or $b \in I$. And this is so, as I is a prime ideal.

We already know that R/I is a commutative ring with unity (see Theorem 3.6, page 123).

(b) Invoking Theorem 3.11, we show that the only ideals of R/I are $\{0\}$ and R/I:

Let I be a maximal ideal in R, and let $\overline{K}$ be an ideal in R/I. Exercise 18, page 126, assures us that there exists an ideal K in R with $I \subseteq K \subseteq R$ such that $\overline{K} = K/I$. Since I is a maximal ideal, either $K = I$, in which case $\overline{K} = \{0\}$, or $K = R$, in which case $\overline{K} = R/I$.

3.3 Integral Domains and Fields

FIELDS OF QUOTIENTS

Let's mimic the development in which the integers Z blossom into the field of rational numbers Q, to one that nurtures a general integral domain D into its field of quotients F:

From Z to Q	From D to the field of quotients F
Let $S_Z = \left\{ \dfrac{a}{b} \middle\| a, b \in Z, \text{ with } (b \neq 0) \right\}$.	Let $S_D = \{(a, b) \| a, b \in D, \text{ with } (b \neq 0)\}$.
In Example 1.9, page 29, we demonstrated that the relation $\dfrac{a}{b} \sim \dfrac{c}{d}$ if $ad = bc$ is an equivalence relation on S_Z.	Following the procedure of Example 1.9, page 29, one can show that the relation $(a, b) \sim (c, d)$ if $ad = bc$ is an equivalence relation on S_D.
Let Q denote the set of equivalent classes associated with the above equivalence relation on S_Z.	Let F denote the set of equivalent classes associated with the above equivalence relation on S_D.
Define addition and multiplication in Q as follows: $\left[\dfrac{a}{b}\right] + \left[\dfrac{c}{d}\right] = \left[\dfrac{a+c}{bd}\right]$ and $\left[\dfrac{a}{b}\right]\left[\dfrac{c}{d}\right] = \left[\dfrac{ac}{bd}\right]$	Define addition and multiplication in F as follows: $[(a, b)] + [(c, d)] = [(ad + cb, bd)]$ and $[(a, b)][(c, d)] = [(ac, bd)]$
You are invited to show in the exercises that the above operations are well defined; which is to say: If $\dfrac{a}{b} \sim \dfrac{a'}{b'}$ and $\dfrac{c}{d} \sim \dfrac{c'}{d'}$, then: $\dfrac{a+c}{bd} \sim \dfrac{a'+c'}{b'd'}$ and $\dfrac{ac}{bd} \sim \dfrac{a'c'}{b'd'}$	You are invited to show in the exercises that the above operations are well defined; which is to say: If $(a, b) \sim (a', b')$ and $(c, d) \sim (c', d')$, then: $(ad + cb, bd) \sim (a'd' + c'b', b'd')$ and $(ac, bd) \sim (a'c', b'd')$
You are also invited to show in the exercises that: $\langle Q, +, \cdot \rangle$ is a field, with zero $\left[\dfrac{0}{1}\right]$ and unity $\left[\dfrac{1}{1}\right]$.	You are also invited to show in the exercises that: $\langle F, +, \cdot \rangle$ is a field, with zero $[(0, a)]$ and unity $[(a, a)]$, for any $a \neq 0$.

As is the case with the rational numbers, where the equivalence class $\left[\dfrac{a}{b}\right]$ is simply denoted by the "fraction" $\dfrac{a}{b}$, so then one generally represents an element $[(a, b)]$ in the field of quotients F by the two-tuple (a, b).

EXERCISES

Exercise 1-6. Find the zero-divisors of the given ring.

1. $3Z$
2. Z_4
3. $Z_3 \times Z_6$
4. $Z \times Z_5$
5. $Z_2 \times Z_5$
6. $Z_4 \times Z_8$

Exercise 7-12. Determine the characteristic of the given ring.

7. $3Z$
8. Z_4
9. $Z_3 \times Z_6$
10. $Z \times Z_5$
11. $Z_2 \times Z_5$
12. $Z_4 \times Z_8$

Exercise 13-15. Solve the equation $x^2 - 5x + 6 = 0$ in:

13. Z_2
14. Z_5
15. Z_{12}

Exercise 16-18. Solve the equation $x^3 - 3x - 4 = 0$ in:

16. Z_2
17. Z_5
18. Z_{12}

19. Show that Z_p has no zero divisors for any prime p.
20. Show that the zero divisors of Z_n are the nonzero elements that are not relatively prime to n.
21. Show that every nonzero element in Z_n is a unit or a zero-divisor.
22. Let R be a finite commutative ring with unity. Prove that every nonzero element in Z_n is a unit or a zero-divisor.
23. Give an example of a ring R that contains a nonzero element that is neither a zero-divisor nor a unit.
24. Show that any nonzero element a in a commutative ring R is a zero-divisor if and only if $a^2 b = 0$ for some $b \neq 0$.
25. Let R and S be nonzero rings. Can $R \times S$ be an integral domain?
26. Give an example of a commutative ring R without zero-divisors that is not an integral domain.
27. A nonempty subset S of an integral domain D is called a **subdomain** of D if it is an integral domain under the operations of D. Prove that a nonempty subset of D is a subdomain of D if and only if S is a subring of D that contains the unity of D.
28. Prove that the intersection of two subdomains of an integral domain D is also a subdomain of D. (See Exercise 18.)
29. Find all subdomains of Z. (See Exercise 27.)

30. Show that the only subdomain of Z_p, for p prime, is Z_p. (See Exercise 18.)

31. Let D be an integral domain of prime characteristic p. Show that for every $a, b \in D$:
$$(a+b)^p = a^p + b^p$$

32. Prove that every maximal ideal in a commutative ring with identity is a prime ideal.

33. Let I be an ideal in a commutative ring R with unity. Prove that I is a prime ideal of R if and only if R/I is an integral domain. [See Theorem 3.15(a).]

34. Let I be an ideal in a commutative ring R with unity. Prove that I is maximal in R if and only if R/I is afield. [See Theorem 3.15(b).]

35. Prove that every proper ideal on a ring with unity is contained in a maximal ideal.

36. Let R be a commutative ring. Prove that if P is a prime ideal of R that contains no zero-divisors, then R is an integral domain.

37. Let R be a commutative ring. Let I and J be ideals of R. Show that if P is a prime ideal of R that contains $I \cap J$, then either I or J is contained in P.

38. Show that the subset $S = \{0, 3\}$ is an ideal in Z_6. Show that while S is not an integral domain, Z_6/S is a field.

39. Show that any ring homomorphism $\phi: F \to R$ from a field F to a ring $R \neq \{0\}$ is one-to-one.

40. Let R be a commutative ring. Prove that R is a field if and only if $\{0\}$ is a maximal ideal.

41. Referring to the "From D to the field of quotients F" development on page 133, verify that the operations:
$$[(a, b)] + [(c, d)] = [(ad+cb, bd)] \text{ and } [(a, b)][(c, d)] = [(ac, bd)]$$
are well defined.

42. Referring to the "From Z to Q" development on page 133, verify that the operations:
$$\left[\frac{a}{b}\right] + \left[\frac{c}{d}\right] = \left[\frac{a+c}{bd}\right] \text{ and } \left[\frac{a}{b}\right]\left[\frac{c}{d}\right] = \left[\frac{ac}{bd}\right]$$
are well defined.

43. Referring to the "From Z to Q" development on page 133, verify that $\langle Q, +, \cdot \rangle$ is a field, with zero $\left[\frac{0}{1}\right]$ and unity $\left[\frac{1}{1}\right]$.

44. Referring to the "From D to the field of quotients F" development on page 133, verify that $\langle F, +, \cdot \rangle$ is a field, with zero $[(0, a)]$ and unity $[(a, a)]$, for any $a \neq 0$.

45. Establish **Fermat's Little Theorem**: If $a \in Z$ and if p is a prime not dividing p, then:
$$a^{p-1} \equiv 1 \pmod{p}$$

46. Show that for any prime p and any $a \in Z$: $a^p \equiv a \pmod{p}$

| | **PROVE OR GIVE A COUNTEREXAMPLE** | |

47. The intersection of subdomains of an integral domain D is a subdomain of D. (See Exercise 18.)

48. If $\phi: D \to R$ is a homomorphism from the integral domain D to a ring R, then $\phi(D)$ is an integral domain.

49. Let R be a commutative ring with unity. If P is a prime ideal of R and if J is a subring of R, then $P \cap J$ is a prime ideal of R.

50. Let R be a commutative ring with unity. If P is a prime ideal of R and if I is an ideal of R, then $P \cap J$ is a prime ideal of R.

APPENDIX A
CHECK YOUR UNDERSTANDING SOLUTIONS
PART 1
PRELIMINARIES

1.1 Functions

CYU 1.1 For $f: M_{2 \times 2} \to \Re$, given by $f\left(\begin{bmatrix} a & b \\ c & d \end{bmatrix}\right) = a + d$, and $g: \Re \to R^2$ given by $g(x) = (2x, x^2)$, we have:

(a) $(g \circ f)\left(\begin{bmatrix} 1 & 3 \\ 2 & 4 \end{bmatrix}\right) = g\left[f\left(\begin{bmatrix} 1 & 3 \\ 2 & 4 \end{bmatrix}\right)\right] = g(1+4) = g(5) = (2 \cdot 5, 5^2) = (10, 25)$

(b) $(g \circ f)\left(\begin{bmatrix} a & b \\ c & d \end{bmatrix}\right) = g\left[f\left(\begin{bmatrix} a & b \\ c & d \end{bmatrix}\right)\right] = g(a+d) = (2(a+d), (a+d)^2)$
$= (2a + 2d, a^2 + 2ad + b^2)$

CYU 1.2 (a) Let $f: M_{2 \times 2} \to R^4$ be given by $f\left(\begin{bmatrix} a & b \\ c & d \end{bmatrix}\right) = (d, -c, 3a, b)$

One-to-one: $f\left(\begin{bmatrix} a & b \\ c & d \end{bmatrix}\right) = f\left(\begin{bmatrix} \bar{a} & \bar{b} \\ \bar{c} & \bar{d} \end{bmatrix}\right) \Rightarrow (d, -c, 3a, b) = (\bar{d}, -\bar{c}, 3\bar{a}, \bar{b})$

$\Rightarrow \left.\begin{array}{l} d = \bar{d} \\ -c = -\bar{c} \\ 3a = 3\bar{a} \\ b = \bar{b} \end{array}\right\} \Rightarrow \left.\begin{array}{l} d = \bar{d} \\ c = \bar{c} \\ a = \bar{a} \\ b = \bar{b} \end{array}\right\} \Rightarrow \begin{bmatrix} a & b \\ c & d \end{bmatrix} = \begin{bmatrix} \bar{a} & \bar{b} \\ \bar{c} & \bar{d} \end{bmatrix}$

Onto: For given (x, y, z, w), we find $\begin{bmatrix} a & b \\ c & d \end{bmatrix}$ such that $f\left(\begin{bmatrix} a & b \\ c & d \end{bmatrix}\right) = (x, y, z, w)$:

$f\left(\begin{bmatrix} a & b \\ c & d \end{bmatrix}\right) = (x, y, z, w) \Rightarrow (d, -c, 3a, b) = (x, y, z, w) \Rightarrow \left.\begin{array}{l} d = x \\ -c = y \\ 3a = z \\ b = w \end{array}\right\} \Rightarrow \left.\begin{array}{l} d = x \\ c = -y \\ a = z/3 \\ b = w \end{array}\right\}$

Hence: $f\left(\begin{bmatrix} z/3 & w \\ -y & x \end{bmatrix}\right) = (x, y, z, w)$.

(b) $f\left(\begin{bmatrix} a & b \\ c & d \end{bmatrix}\right) = \begin{bmatrix} b & a \\ c+d & 2b \end{bmatrix}$ is not one-to-one: $f\left(\begin{bmatrix} 0 & 0 \\ 0 & 0 \end{bmatrix}\right) = f\left(\begin{bmatrix} 0 & 0 \\ 1 & 0 \end{bmatrix}\right) = \begin{bmatrix} 0 & 0 \\ 0 & 0 \end{bmatrix}$.

f is not onto, since no element $\begin{bmatrix} a & b \\ c & d \end{bmatrix}$ is mapped to $\begin{bmatrix} 1 & 0 \\ 0 & 3 \end{bmatrix}$: $f\left(\begin{bmatrix} a & b \\ c & d \end{bmatrix}\right) = \begin{bmatrix} b & a \\ c+d & 2b \end{bmatrix} \neq \begin{bmatrix} 1 & 0 \\ 0 & 3 \end{bmatrix}$.

CYU 1.3 Let $y \in Y$. Since $[y, f^{-1}(y)] \in f^{-1}$, $[f^{-1}(y), y] \in f$, which is to say: $f[f^{-1}(y)] = y$.

CYU 1.4 The function $f: M_{2 \times 2} \to R^4$ given by $f\left(\begin{bmatrix} a & b \\ c & d \end{bmatrix}\right) = (d, -c, 3a, b)$ is a bijection [see CYU 1.2(a)]. To find its inverse we determine $\begin{bmatrix} a & b \\ c & d \end{bmatrix}$ for which $f\left(\begin{bmatrix} a & b \\ c & d \end{bmatrix}\right) = (x, y, z, w)$:

$$f\left(\begin{bmatrix} a & b \\ c & d \end{bmatrix}\right) = (x, y, z, w) \Rightarrow (d, -c, 3a, b) = (x, y, z, w) \Rightarrow \left.\begin{matrix} d = x \\ -c = y \\ 3a = z \\ b = w \end{matrix}\right\} \Rightarrow \left.\begin{matrix} a = z/3 \\ b = w \\ c = -y \\ d = x \end{matrix}\right\}$$

Conclusion: $f^{-1}(x, y, z, w) = \begin{bmatrix} z/3 & w \\ -y & x \end{bmatrix}$

Moreover: $f[f^{-1}(x, y, z, w)] = f\left(\begin{bmatrix} z/3 & w \\ -y & x \end{bmatrix}\right) = \left(x, -(-y), 3\left(\frac{z}{3}\right), w\right) = (x, y, z, w)$

and: $f^{-1}\left[f\left(\begin{bmatrix} a & b \\ c & d \end{bmatrix}\right)\right] = f^{-1}(d, -c, 3a, b) = \begin{bmatrix} 3(a/3) & b \\ -(-c) & d \end{bmatrix} = \begin{bmatrix} a & b \\ c & d \end{bmatrix}$

CYU 1.5 From $f(x, y, z, w) = \begin{bmatrix} -y & 2x \\ 3w & z \end{bmatrix}$ and $g\left(\begin{bmatrix} a & b \\ c & d \end{bmatrix}\right) = (d, -c, 3a, b)$ we have:

$(g \circ f)(x, y, z, w) = g[f(x, y, z, w)] = g\left(\begin{bmatrix} -y & 2x \\ 3w & z \end{bmatrix}\right) = (z, -3w, -3y, 2x)$

To find its inverse of $g \circ f: \Re^4 \to \Re^4$ we start with $(x, y, z, w) \in \Re^4$ on the right side of $g \circ f: \Re^4 \to \Re^4$ and find (a, b, c, d) (on the left side) for which $(g \circ f)(a, b, c, d) = (x, y, z, w)$ (we will then turn things around to arrive at $(g \circ f)^{-1}$). Let's do it:

$$(g \circ f)(a, b, c, d) = (x, y, z, w) \Rightarrow g[f(a, b, c, d)] = (x, y, z, w)$$

$$\Rightarrow g\left(\begin{bmatrix} -b & 2a \\ 3d & c \end{bmatrix}\right) = (x, y, z, w) \Rightarrow (c, -3d, -3b, 2a) = (x, y, z, w)$$

$$\Rightarrow \begin{array}{l} c = x \\ -3d = y \\ -3b = z \\ 2a = w \end{array} \Rightarrow \begin{array}{l} a = w/2 \\ b = -z/3 \\ c = x \\ d = -y/3 \end{array}$$

At this point we have $(g \circ f)\left(\dfrac{w}{2}, -\dfrac{z}{3}, x, -\dfrac{y}{3}\right) = (x, y, z, w)$; and, consequently:

$$(g \circ f)^{-1}(x, y, z, w) = \left(\dfrac{w}{2}, -\dfrac{z}{3}, x, -\dfrac{y}{3}\right)$$

We now verify that $(f^{-1} \circ g^{-1})(x, y, z, w)$ also equals $\dfrac{w}{2}, -\dfrac{z}{3}, x, -\dfrac{y}{3}$, where

$$g^{-1}(x, y, z, w) = \begin{bmatrix} z/3 & w \\ -y & x \end{bmatrix} \text{ and } f^{-1}\left(\begin{bmatrix} a & b \\ c & d \end{bmatrix}\right) = \left(\dfrac{b}{2}, -a, d, \dfrac{c}{3}\right):$$

$$(f^{-1} \circ g^{-1})(x, y, z, w) = f^{-1}[g^{-1}(x, y, z, w)] = f^{-1}\left(\begin{bmatrix} z/3 & w \\ -y & x \end{bmatrix}\right) = \left(\dfrac{w}{2}, -\dfrac{z}{3}, x, -\dfrac{y}{3}\right)$$

1.2 Principle of Mathematical Induction

CYU 1.6 (a) The equation $2 + 4 = 1 + 2 + 3 + 4 - (1 + 3)$ illustrate that the sum of the first two even integers can be expressed as the sum of the first four integers minus the sum of the first two odd integer. Generalizing, we anticipate that the sum of the first n even integers is the sum of the first **2n integers** minus the sum of the first **n odd integers**; leading us to the conjecture that the sum of the first n even integers equals $n^2 + n$:

$$\underbrace{\dfrac{2n(2n+1)}{2}}_{\substack{\text{sum or first 2n integers} \\ \text{(Eample 1.16)}}} - \underbrace{n^2}_{\substack{\text{sum of first } n \text{ odd integers} \\ \text{(page 34)}}} = 2n^2 + n - n^2 = n^2 + n$$

(b) Let $P(n)$ be the proposition that the sum of the first n even integers equals $n^2 + n$.

I. Since the sum of the first 1 even integers is 2, $P(1) = 1^2 + 1 = 2$ is true.

II. Assume $P(k)$ is true; that is: $\mathbf{2 + 4 + 6 + \cdots + 2k = k^2 + k}$.

III. We complete the proof by verifying that $P(k+1)$ is true; which is to say, that $2 + 4 + 6 + \cdots + 2k + 2(k+1) = (k+1)^2 + (k+1)$:

$$\underbrace{2 + 4 + 6 + \cdots + 2k}_{\text{by II}} + 2(k+1) = k^2 + k + 2(k+1)$$

$$= (k^2 + 2k + 1) + (k+1) = (k+1)^2 + (k+1)$$

CYU 1.7 (a) False — a counterexample: $4|(3+1)$, and 4 divides neither 3 nor 1.

(b) True: Since $a|b$, there exists h such that: (1) $b = ah$.
Since $a|(b+c)$, there exists k such that: (2) $b+c = ak$.
From (2): $c = ak - b$. From (1): $c = ak - ah = a(k-h)$.
Since $c = at$ (where $t = k-h$): $a|c$.

CYU 1.8 (a) Let $P(n)$ be the proposition $n! > n^2$:

I. $P(4)$ is true: $4! = 1 \cdot 2 \cdot 3 \cdot 4 = 24 \geq 4^2$.

II. **Assume** $P(k)$ is true: $k! > k^2$ (for $k \geq 4$)

III. We show $P(k+1)$ is true; namely, that $(k+1)! > (k+1)^2$:

$$(k+1)! = k!(k+1) > k^2(k+1)$$
$$\text{II} \uparrow$$

Now what? Well, if we can show that $k^2(k+1) > (k+1)^2$, then we will be done. Let's do it:

Since $k \geq 4$, $k \geq 2$, and therefore $k^2 = k \cdot k > 2k > k+1$.

Multiplying both sides by the positive number $(k+1)$: $k^2(k+1) > (k+1)^2$.

(b) Let $P(n)$ be the proposition that $6|(n^3 + 5n)$ for all integers $n \geq 1$.

I. True at $n = 1$: $6|(1^3 + 5 \cdot 1)$.

II. Assume $P(k)$ is true; that is: $6|(k^3 + 5k)$.

III. To establish that $6|[(k+1)^3 + 5(k+1)]$, we begin by noting that $(k+1)^3 + 5(k+1) = (k^3 + 3k^2 + 8k) + \mathbf{6}$ and then set our sights on showing that $6|(k^3 + 3k^2 + 8k)$ (for clearly $6|6$).

Wanting to get II into play we rewrite $k^3 + 3k^2 + 8k$ in the form $(k^3 + 5k) + (\mathbf{3k^2 + 3k})$. Our induction hypothesis allows us to assume that $6|(k^3 + 5k)$. If we can show that $6|(\mathbf{3k^2 + 3k})$, then we will be done, by virtue of Theorem 1.6(b), page 28. Let's do it:

Since $3k^2 + 3k = 3k(k+1)$, and since either k or $k+1$ is even:

6 is a factor of $3x^2 + 3k$.

CYU 1.9 Let $P(n)$ be a proposition for which $P(1)$ is True and for which the validity at k implies the validity at $k+1$. We are to show, using the Well-Ordering Principle, that $P(n)$ is True for all n. Suppose not (we will arrive at a contradiction):

Let $S = \{n \in Z^+ | P(n) \text{ is False}\}$. Since $P(1)$ is True, $S \neq \emptyset$. The Well-Ordering Principle tells us that S contains a least element, n_0. But since the validity at $n_0 - 1$ implies the validity at n_0, $n_0 - 1$ must be in S — contradicting the minimality of n_0.

1.3 The Division Algorithm and Beyond

CYU 1.10 The division algorithm tells us that n must be of the form $3m$, or $3m+1$, or $3m+2$, for some integer m. We show that, in each case, $n^2 = 3q$ or $n^2 = 3q+1$ for some integer q:

If $n = 3m$, then $n^2 = 9m^2 = 3q$ with $q = 3m^2$.

If $n = 3m+1$, then $n^2 = 9m^2 + 6m + 1 = 3(3m^2 + 2m) + 1 = 3q+1$.

If $n = 3m+2$, then $n^2 = 9m^2 + 12m + 4 = 3(3m^2 + 4m + 1) + 1 = 3q+1$.

CYU 1.11 We simply show that $c > 0$ divides $n \in Z$ if and only if $c||n|$:
$$c = kn \Leftrightarrow |c| = |kn| \Leftrightarrow |c| = |k||n| \underset{\text{since } c > 0}{\Leftrightarrow} c = h|n| \text{ where } h = |k|$$

CYU 1.12 Proof by contradiction: Assume that $gcd(a, c) = 1$. From Theorem 1.9: if $a|bc$, and if $gcd(a, c) = 1$, then $a|b$ — contradicting the given condition that $a \nmid b$.

CYU 1.13 Let $P(n)$ be the proposition that if $p|a_1 a_2 \cdots a_n$, then $p|a_i$ for some $1 \leq i \leq n$.

    I. $P(1)$ is trivially True.

    II. Assume $P(k)$ is True: If $p|a_1 a_2 \cdots a_k$, then $p|a_i$ for some $1 \leq i \leq k$.

    III. Suppose $p|a_1 a_2 \cdots a_k a_{k+1}$; or, to write it another way: $p|(a_1 a_2 \cdots a_k)a_{k+1}$. If $p|a_{k+1}$ then we are done. If not, then by Theorem 1.8: $p|(a_1 a_2 \cdots a_k)$. Invoking II we conclude that $p|a_i$ for some $1 \leq i \leq k$.

CYU 1.14 Let $a = p_1^{r_1} p_2^{r_2} \ldots p_s^{r_s}$, $b = q_1^{m_1} q_2^{m_2} \ldots q_t^{m_t}$ be the prime decompositions of a and b, with distinct primes $p_1, p_2, \ldots, p_s$, and distinct primes $q_1, q_2, \ldots, q_t$.

Since $a|n$: $n = ak = p_1^{r_1} p_2^{r_2} \ldots p_s^{r_s} \cdot k$. It follows that and each $p_i^{r_i}$ must appear in the prime decomposition of n, for $1 \leq i \leq s$ (with possibly additional p_j's appearing in the prime decomposition of k). Similarly, since $b|n$, each $q_i^{m_i}$ must appear in the prime decomposition of n, for $1 \leq i \leq t$.

Since a and b are relatively prime, none of the p_i's is equal to any of the q_i's. It follows that $p_1^{r_1} p_2^{r_2} \ldots p_s^{r_s} q_1^{m_1} q_2^{m_2} \ldots q_t^{m_t}$ appears in the prime decomposition of n, and that therefore $ab = p_1^{r_1} p_2^{r_2} \ldots p_s^{r_s} q_1^{m_1} q_2^{m_2} \ldots q_t^{m_t}$ divides n.

A-6   APPENDIX A

1.4 Equivalence Relations

CYU 1.15 **Reflexive:** Let $A \in S$. Since $I: A \to A$, given by $I(a) = a \; \forall a \in A$ is a bijection, $A \sim a$.

Symmetric: If $A \sim B$ for $A, B \in S$, then there exists a bijection $f: A \to B$. Theorem 1.1(a), page 5, tells us that $f^{-1}: B \to A$ is a bijection. Hence, $B \sim A$.

Transitive: If $A \sim B$ and $B \sim C$ with $A, B, C \in S$, then there exists bijections $f: A \to B$ and $g: B \to C$. Theorem 1.2(c), page 7, tells us that $g \circ f: A \to C$ is a bijection. Hence, $A \sim C$.

CYU 1.16 (a) No: $[1, 2] \cap [2, 3] \neq \varnothing$.

(b) Yes: Every element of $\Re$ is either an integer or is contained in some $(i, i+1)$ for some integer $i \geq 0$ or in some $(-i, -i-1)$ for some $i \geq 1$. Moreover the sets in $\{\{n\} | n \in Z\} \cup \{(i, i+1)\}_{i=0}^{\infty} \cup \{(-i, -i-1)\}_{i=1}^{\infty}$ are mutually disjoint.

CYU 1.17 Let $a = d_a n + r_a$ and $b = d_b n + r_b$ with $0 \leq r_a < n$ and $0 \leq r_b < n$.

If $r_a = r_b$, then $a - b = d_a n - d_b n = n(d_a - d_b)$. Since $n | (a-b)$, $a \equiv b \bmod n$.

Conversely, assume that $r_a \neq r_b$, say: $0 \leq r_b < r_a < n$. Then:

$$a - b = (d_a n + r_a) - (d_b n + r_b) = (d_a - d_b)n + (r_a - r_b)$$

Assume that $n | (a - b)$. Since $r_a - r_b = (a - b) - (d_a - d_b)n$, n would have to divide $(r_a - r_b)$; which it cant, sine $0 < r_a - r_b < n$. It follows that $r_a \neq r_b \Rightarrow a \not\equiv b \bmod n$.

CYU 1.18 (a) $[a]_n = [\bar{a}]_n \Rightarrow (a - \bar{a}) = hn$ and $[b]_n = [\bar{b}]_n \Rightarrow (b - \bar{b}) = kn$, for $h, k \in Z$.

Then: $ab - \bar{a}\bar{b} = (hn + \bar{a})b - \bar{a}(b - kn) = hbn + \bar{a}b - \bar{a}b + k\bar{a}n = (hb + k\bar{a})n$

Since $n | (ab - \bar{a}\bar{b})$, $[ab]_n = [\bar{a}\bar{b}]_n$.

(b) $[a]_n([b]_n[c]_n) = [a]_n([bc]_n) = [a(bc)]_n$
$= [(ab)c]_n = [ab]_n[c]_n = ([a]_n[b]_n)[c]_n$

(c) $[a]_n([b]_n + [c]_n) = [a]_n([b+c]_n) = [a(b+c)]_n$
$= [ab + ac]_n = [ab]_n + [ac]_n = [a]_n[b]_n + [a]_n[c]_n$

PART 2
GROUPS

2.1 Definitions and Examples

CYU 2.1 (a) Closure: The sum of two n-tuples is again an n-tuple.

Associative:
$$[(a_1, a_2, ..., a_n) + (b_1, b_2, ..., b)] + (c_1, c_2, ..., c_n) = (a_1 + b_1, a_2 + b_2, ..., a_n + b_n) + (c_1, c_2, ..., c_n)$$
$$= (a_1 + b_1, a_2 + b_2, ..., a_n + b_n) + (c_1, c_2, ..., c_n)$$
$$= ([a_1 + b_1] + c_1, [a_2 + b_2] + c_2, ..., [a_n + b_n] + c_n)$$
$$= (a_1 + [b_1 + c_1], a_2 + [b_2 + c_2], ..., a_n + [b_n + c_n])$$
$$= (a_1, a_2, ..., a_n) + [(b_1, b_2, ..., b) + (c_1, c_2, ..., c_n)]$$

Identity: $(a_1, a_2, ..., a_n) + (0, 0, ..., 0) = (a_1 + 0, a_2 + 0, ..., a_n + 0) = (a_1, a_2, ..., a_n)$

Invere: $(a_1, a_2, ..., a_n) + (-a_1, -a_2, ..., -a_n) = (a_1 + (-a_1), a_2 + (-a\ ), ..., a_n + (-a_n))$
$$= (0, 0, ..., 0)$$

(b) Closure: The sum of 2 two-by-two matrices is again a two-by-two matrix.

Associative:
$$\left\{\begin{bmatrix} a_1 & b_1 \\ c_1 & d_1 \end{bmatrix} + \begin{bmatrix} a_2 & b_2 \\ c_2 & d_2 \end{bmatrix}\right\} + \begin{bmatrix} a_3 & b_3 \\ c_3 & d_3 \end{bmatrix} = \begin{bmatrix} a_1 + a_2 & b_1 + b_2 \\ c_1 + c_2 & d_1 + d_2 \end{bmatrix} + \begin{bmatrix} a_3 & b_3 \\ c_3 & d_3 \end{bmatrix} = \begin{bmatrix} (a_1 + a_2) + a_3 & (b_1 + b_2) + b_3 \\ (c_1 + c_2) + c_3 & (d_1 + d_2) + d_3 \end{bmatrix}$$
$$= \begin{bmatrix} a_1 + (a_2 + a_3) & b_1 + (b_2 + b_3) \\ c_1 + (c + c_3) & d_1 + (d_2 + d_3) \end{bmatrix}$$
$$= \begin{bmatrix} a_1 & b_1 \\ c_1 & d_1 \end{bmatrix} + \left\{\begin{bmatrix} a_2 & b_2 \\ c_2 & d_2 \end{bmatrix} + \begin{bmatrix} a_3 & b_3 \\ c_3 & d_3 \end{bmatrix}\right\}$$

Identity: $\begin{bmatrix} a & b \\ c & d \end{bmatrix} + \begin{bmatrix} 0 & 0 \\ 0 & 0 \end{bmatrix} = \begin{bmatrix} a+0 & b+0 \\ c+0 & d+0 \end{bmatrix} = \begin{bmatrix} a & b \\ c & d \end{bmatrix}$

Inverse: $\begin{bmatrix} a & b \\ c & d \end{bmatrix} + \begin{bmatrix} -a & -b \\ -c & -d \end{bmatrix} = \begin{bmatrix} a+(-a) & b+(-b) \\ c+-c) & d+(-d) \end{bmatrix} = \begin{bmatrix} 0 & 0 \\ 0 & 0 \end{bmatrix}$

CYU 2.2 The values in column a follow from the observation that $0 +_n n = n$ for $0 \leq n \leq 3$.

As for column b, row 3: $3 +_4 1 = 0$, since $3 + 1 = 4 = 1 \cdot 4 + 0$

As for column c, rows 2 and 3: $2 +_4 2 = 0$ and $3 +_4 2 = 1$, since:
$2 + 2 = 4 = 1 \cdot 4 + 0$ and $3 + 2 = 5 = 1 \cdot 4 + 1$.

As for column d, rows 1, 2, and 3: $1 +_4 3 = 0$, $2 +_4 3 = 1$,
and $3 +_4 3 = 2$, since: $1 + 3 = 4 = 1 \cdot 4 + 0$,
$2 + 3 = 5 = 1 \cdot 4 + 1$, $3 + 3 = 6 = 1 \cdot 4 + 2$.

	a	b	c	d
$+_4$	0	1	2	3
0	0	1	2	3
1	1	2	3	0
2	2	3	0	1
3	3	0	1	2

CYU 2.3 Let $G = \{e, a_1, a_2, \ldots, a_{n-1}\}$. By construction, the ith column of G's group table is precisely $ea_i, a_1a_i, a_2a_i, \ldots, a_{n-1}a_i$. The fact that every element of G appears exactly one time in that row is a consequence of Exercise 37, which asserts that the function $k_{a_i}: G \to G$ given by $k_{a_i}(g) = ga_i$ is a bijection.

CYU 2.4 From $\sigma = \begin{pmatrix} 1 & 2 & 3 & 4 & 5 \\ 1 & 5 & 2 & 3 & 4 \end{pmatrix}$ and $\tau = \begin{pmatrix} 1 & 2 & 3 & 4 & 5 \\ 5 & 3 & 2 & 1 & 4 \end{pmatrix}$ we have:

$$\tau \circ \sigma \underset{\downarrow \tau}{\overset{\sigma}{\longrightarrow}} \begin{pmatrix} 1 & 2 & 3 & 4 & 5 \\ 1 & 5 & 2 & 3 & 4 \\ 5 & 4 & 3 & 2 & 1 \end{pmatrix} \Rightarrow \tau \circ \sigma : \begin{pmatrix} 1 & 2 & 3 & 4 & 5 \\ 5 & 4 & 3 & 2 & 1 \end{pmatrix}$$

$$\sigma \circ \tau \underset{\downarrow \sigma}{\overset{\tau}{\longrightarrow}} \begin{pmatrix} 1 & 2 & 3 & 4 & 5 \\ 5 & 3 & 2 & 1 & 4 \\ 4 & 2 & 5 & 1 & 3 \end{pmatrix} \Rightarrow \sigma \circ \tau : \begin{pmatrix} 1 & 2 & 3 & 4 & 5 \\ 4 & 2 & 5 & 1 & 3 \end{pmatrix}$$

CYU 2.5 (a) We know that 1 and 5 are generators of Z_6 [Example 2.2(a)]. The remaining 4 elements in Z_6 are not:

$$0 +_6 0 = 0 \qquad 2 +_6 2 = 4 \qquad 3 +_6 3 = 0 \qquad 4 +_6 4 = 2$$
$$\qquad\qquad\quad 2 +_6 2 +_6 2 = 0 \qquad\qquad\quad 4 +_6 4 +_6 4 = 0$$

(b) $S_2: \begin{cases} \overset{\alpha_0}{1 \to 1} & \overset{\alpha_1}{1 \to 2} \\ 2 \to 2 & 2 \to 1 \end{cases}$ is cyclic, with generator $\alpha_1: \alpha_1 \circ \alpha_1 = \begin{matrix} 1 \to 2 \to 1 \\ 2 \to 1 \to 2 \end{matrix} = \alpha_0$.

(c) For $n > 2$, consider the following bijections $\beta, \gamma \in S_n$:

$\beta(1) = 2, \beta(2) = 1$, and $\beta(i) = i$ for $3 \leq i \leq n$

$\gamma(2) = 3, \gamma(3) = 2$, and $\beta(i) = i$ for i not equal to 2 or 3

Since $(\gamma \circ \beta)(1) = \gamma[\beta(1)] = \gamma(2) = 3$ and $(\beta \circ \gamma)(1) = \beta[\gamma(1)] = \beta(1) = 2$, S_n is not abelian, and therefore not cyclic.

2.2 Elementary Properties of Groups

CYU 2.6 Since $(bca)(bca) = (bc)(abc)a = (bc)e(a) = bca$: $bca \cdot bca = e$ (Theorem 2.6).

CYU 2.7 (a) False: For $\beta, \gamma, \delta \in S_3$ given by $\beta: \begin{array}{l} 1 \to 2 \\ 2 \to 3 \\ 3 \to 1 \end{array}$, $\gamma: \begin{array}{l} 1 \to 2 \\ 2 \to 1 \\ 3 \to 3 \end{array}$ and $\delta: \begin{array}{l} 1 \to 3 \\ 2 \to 2 \\ 3 \to 1 \end{array}$ we have:

$\gamma \circ \beta: \begin{array}{l} 1 \xrightarrow{\beta} 2 \xrightarrow{\gamma} 1 \\ 2 \to 3 \to 3 \\ 3 \to 1 \to 2 \end{array}$ and $\beta \circ \delta: \begin{array}{l} 1 \xrightarrow{\delta} 3 \xrightarrow{\beta} 1 \\ 2 \to 2 \to 3 \\ 3 \to 1 \to 2 \end{array}$ as well.

(b) True: $a + b = b + c \underset{\text{commutativity}}{\Rightarrow} b + a = b + c \underset{\text{Theorem 2.9}}{\Rightarrow} a = c$

CYU 2.8 We show that the equation $a + x = b$ has a unique solution in $\langle \Re, + \rangle$:

Existence: $a + x = b \Rightarrow -a + (a + x) = -a + b \Rightarrow (-a + a) + x = -a + b$

$\Rightarrow 0 + x = -a + b \Rightarrow x = -a + b$

Uniqueness: If x and $\bar{x}$ then: $a + x = b$ and $a + \bar{x} = b \Rightarrow a + x = a + \bar{x} \underset{\text{Theorem 2.8}}{\Rightarrow} x = \bar{x}$

CYU 2.9 We know that we have to consider a non-abelian group, and turn to our friend S_3. Specifically for $\alpha, \beta \in S_3$ given by $\alpha: \begin{array}{l} 1 \to 3 \\ 2 \to 2 \\ 3 \to 1 \end{array}$ and $\beta = \begin{array}{l} 1 \to 3 \\ 2 \to 1 \\ 3 \to 2 \end{array}$ we have: $\alpha^{-1} = \begin{array}{l} 1 \to 3 \\ 2 \to 2 \\ 3 \to 1 \end{array}$,

$\beta^{-1} = \begin{array}{l} 1 \to 2 \\ 2 \to 3 \\ 3 \to 1 \end{array}$, and $\beta \circ \alpha = \begin{array}{l} 1 \xrightarrow{\alpha} 3 \xrightarrow{\beta} 2 \\ 2 \to 2 \to 1 \\ 3 \to 1 \to 3 \end{array} = \begin{array}{l} 1 \to 2 \\ 2 \to 1 \\ 3 \to 3 \end{array}$, so that:

$(\beta \circ \alpha)^{-1} = \begin{array}{l} 1 \to 2 \\ 2 \to 1 \\ 3 \to 3 \end{array}$ while $\beta^{-1} \circ \alpha^{-1} = \begin{array}{l} 1 \xrightarrow{\alpha^{-1}} 3 \xrightarrow{\beta^{-1}} 1 \\ 2 \to 2 \to 3 \\ 3 \to 1 \to 2 \end{array} = \begin{array}{l} 1 \to 1 \\ 2 \to 3 \\ 3 \to 2 \end{array}$

CYU 2.10 I. $(a_n \ldots a_2 a_1)^{-1} = a_1^{-1} a_2^{-1} \ldots a_n^{-1}$ clearly holds for $n = 1$.

II. Assume $(a_k \cdots a_2 a_1)^{-1} = a_1^{-1} a_2^{-1} \ldots a_k^{-1}$. Then:

III. $(a_{k+1} \cdot a_k \cdots a_2 a_1)^{-1} = [a_{k+1} \cdot (a_k \cdots a_2 a_1)]^{-1}$

$\underset{\text{Theorem 2.12:}}{=} (a_k \cdots a_2 a_1)^{-1} a_{k+1}^{-1} \underset{\text{II}}{=} a_1^{-1} a_2^{-1} \ldots a_k^{-1} \cdot a_{k+1}^{-1}$

CYU 2.11 (a)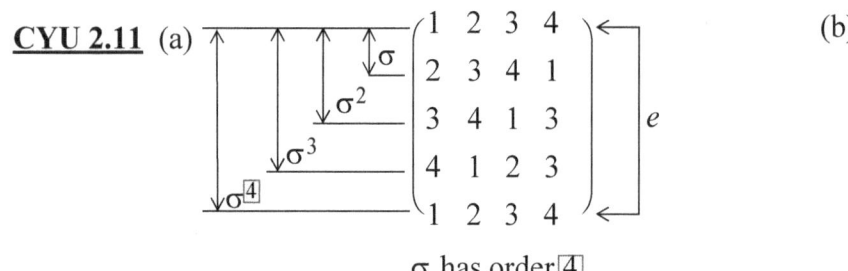

σ has order $\boxed{4}$

(b) $1(4) = 4$
$2(4) = 4 +_{24} 4 = 8$
$3(4) = 8 +_{24} 4 = 12$
$4(4) = 12 +_{24} 4 = 16$
$5(4) = 16 +_{24} 4 = 20$
$6(4) = 20 +_{24} 4 = 0$: 4 has order 6

(c) Let $gcd(a, n) = d$. We show that d is the smallest positive integer m such that $ma = kn$ for some k:

$$ma = kn \Rightarrow m = \frac{kn}{a} = \frac{k\left(\frac{n}{d}\right)}{\frac{a}{d}} \quad (*)$$

Since $gcd(a, n) = d$: $\frac{n}{d}$ and $\frac{a}{d}$ are relatively prime. It follows, from (*), that $\frac{a}{d}\Big| k$. Turning to $m = \frac{kn}{a}$ we see that m will be smallest when k is smallest; which is to say, when $k = \frac{a}{d}$. Hence, the smallest m turns out to be $\frac{kn}{a} = \frac{\frac{a}{d} \cdot n}{a} = \frac{n}{d}$.

2.3 Subgroups

CYU 2.12 We already know that $6Z$ is a subgroup of Z. To show that it is a subgroup of $3Z$ we need but observe that $6Z \subseteq 3Z$: $n \in 6Z \Rightarrow n = 6m$ for $m \in Z$
$$\Rightarrow n = 3(2m) \Rightarrow n \in 3Z$$

CYU 2.13 $h_1 k_1 = h_2 k_2 \Rightarrow h_2^{-1} h_1 = k_2 k_1^{-1} \Rightarrow \begin{cases} h_2^{-1} h_1 = e \Rightarrow h_1 = h_2 \\ k_2 k_1^{-1} = e \Rightarrow k_1 = k_2 \end{cases}$

both in H and K --so

CYU 2.14 We show that $\langle 3 \rangle = Z_8 = \{0, 1, 2, 3, 4, 5, 6, 7\}$ by demonstrating that every element of Z_8 is a multiple of 3:

$1 \cdot 3 = 3, \quad 2 \cdot 3 = 3 +_8 3 = 6, \quad 3 \cdot 3 = 3 +_8 3 +_8 3 = 1, \quad 4 \cdot 3 = 3 +_8 3 +_8 3 +_8 3 = 4$

$5 \cdot 3 = 3 +_8 3 +_8 3 +_8 3 +_8 3 = 7, \quad 6 \cdot 3 = 2, \quad 7 \cdot 3 = 5, \quad 8 \cdot 3 = 0$

Claim: $\langle 4 \rangle = \{0, 4\}$: $1 \cdot 4 = 4$, $2 \cdot 4 = 4 +_8 4 = 0$. Fine, but can we pick up other elements of Z_8 by taking additional multiples of 4? No:

The division algorithm assures us that $n = q4 + r$ for any $n \in Z$, with $0 \le r < 4$. From the above we know that $1 \cdot 4$ and $2 \cdot 4$ are in $\{0, 4\}$, and surely $0 \cdot 4 \in \{0, 4\}$. The only possible loose end is $3 \cdot 4$. Let's tie it up:
$$3 \cdot 4 = (4 +_8 4) +_8 4 = 0 +_8 4 = 4$$

CYU 2.15 A direct consequence of CYU 2.11(c), page 57.

CYU 2.16 $(ii) \Rightarrow (i)$: Let S be the intersection of all subgroups of G that containing A. Since $\langle A \rangle$ is a subgroup of G containing A that is contained in every subgroup of G that contains A: $S = \langle A \rangle$.

CYU 2.17 We show that $\langle \alpha_2, \alpha_3 \rangle = S_3$ by observing that every element in

$$S_3 = \left\{ e = \begin{pmatrix} 1 & 2 & 3 \\ 1 & 2 & 3 \end{pmatrix}, \alpha_1 = \begin{pmatrix} 1 & 2 & 3 \\ 2 & 3 & 1 \end{pmatrix}, \alpha_2 = \begin{pmatrix} 1 & 2 & 3 \\ 3 & 1 & 2 \end{pmatrix}, \alpha_3 = \begin{pmatrix} 1 & 2 & 3 \\ 1 & 3 & 2 \end{pmatrix}, \alpha_4 = \begin{pmatrix} 1 & 2 & 3 \\ 3 & 2 & 1 \end{pmatrix}, \alpha_5 = \begin{pmatrix} 1 & 2 & 3 \\ 2 & 1 & 3 \end{pmatrix} \right\}$$

can be expressed a a product of the permutations α_2, α_3 (details omitted):

$$e = \alpha_3^2, \quad \alpha_1 = \alpha_2^2, \quad \alpha_2 = \alpha_2, \quad \alpha_3 = \alpha_3, \quad \alpha_4 = \alpha_2 \alpha_3, \quad \alpha_5 = \alpha_3 \alpha_2$$

2.4 Homomorphisms and Isomorphisms

CYU 2.18 Let $\phi: G \to G'$ be given by $\phi(a) = e$. Since for every $a, b \in G$, $\phi(ab) = e = ee = \phi(a)\phi(b)$, ϕ is a homomorphism.

CYU 2.19 For $a, b \in G$ we have:

$$(\theta \circ \phi)(a+b) = \theta[\phi(a+b)] = \theta[\phi(a) + \phi(b)] = \theta[\phi(a)] + \theta[\phi(b)]$$

$\uparrow$ —— Definition of composition $\to$ $= (\theta \circ \phi)(a) + (\theta \circ \phi)(b)$

CYU 2.20 Homomorphism:
$$\phi(2n_1 + 2n_2) = \phi[2(n_1 + n_2)] = 8(n_1 + n_2) = 8n_1 + 8n_2 = \phi(2n_1) + \phi(2n_1)$$
$\text{Ker}(\phi) = \{2n | \phi(2n) = 0\} = \{2n | 8n = 0\} = \{0\}$.
$\text{Im}(\phi) = \{\phi(2n)\} = \{8n\} = 8\mathbb{Z}$

CYU 2.21 We are to show that for any $a, b \in G$, $\phi(a) = \phi(b) \Rightarrow a = b$. Let's do it:

$$\phi(a) = \phi(b) \Rightarrow \phi(ca) = \phi(cb) \Rightarrow \phi(c)\phi(a) = \phi(c)\phi(b)$$
$$\Rightarrow \phi(c) = \phi(c)\phi(b)[\phi(a)]^{-1}$$
$$\Rightarrow \phi(c) = \phi(c)\phi(b)\phi(a^{-1})$$
$$\Rightarrow \phi(c) = \phi(cba^{-1}) \Rightarrow c = cba^{-1}$$
$$\Rightarrow c^{-1}c = c^{-1}cba^{-1}$$
$$\Rightarrow e = ba^{-1} \Rightarrow a = b$$

CYU 2.22 (a) We show that the relation $\cong$ given by $G \cong G'$ if G is isomorphic to G' is an equivalence relation:

Reflexive $G \cong G$ since the identity map $I(g) = g$ is clearly an isomorphism.

Symmetric $G \cong G' \Rightarrow G' \cong G$: Let $\phi: G \to G'$ be an isomorphism. Theorem 1.1(a), page 5, assures us that the map $\phi^{-1}: G' \to G$ is a bijection. We show that it is also a homomorphism:

For $a', b' \in G'$ let $a, b \in G$ be such that $\phi(a) = a'$ and $\phi(b) = b'$. Since $\phi(ab) = a'b'$, We then have: $\phi^{-1}(a'b') = ab = \phi^{-1}(a')(\phi^{-1}(b'))$.

Transitive $G_1 \cong G_2$ and $G_2 \cong G_3 \Rightarrow G_1 \cong G_3$: Follows from Theorem 1.2(c), page 7, and CYU 2.19.

(b) We show that the map $\phi: nZ \to mZ$ given by $\varphi(nz) = mz$ is an isomorphism:

One-to one: $\phi(nz) = \phi(n\bar{z}) \Rightarrow mz = m\bar{z} \Rightarrow z = \bar{z}$

Onto: For given $mz \in mZ$: $\phi(nz) = mz$.

Homomorphism:
$$\phi(nz + n\bar{z}) = \phi[n(z+\bar{z})] = m(z+\bar{z}) = mz + m\bar{z} = \phi(nz) + \phi(n\bar{z})$$

(c) Let $g \in G$. The map $i_g: G \to G$ is a bijection:

One-to one: $i_g(a) = i_g(b) \Rightarrow gag^{-1} = gbg^{-1} \Rightarrow g^{-1}gag^{-1}g = g^{-1}gbg^{-1}g \Rightarrow a = b$

Onto: For $a \in G$, $i_g(g^{-1}ag) = g[g^{-1}ag]g^{-1} = a$

Homomorphism: $i_g(ab) = gabg^{-1} = (gag^{-1})(gbg^{-1}) = i_g(a)i_g(b)$

CYU 2.23 We show that $\phi: \langle a \rangle \to Z$ given by $\phi(a^n) = n$ is an isomorphism:

One-to-one: $\phi(a^n) = \phi(a^m) \Rightarrow n = m \Rightarrow a^n = a^m$

Onto: For $n \in Z$, $\phi(a^n) = n$.

Homomorphism: $\phi(a^n a^m) = \phi(a^{n+m}) = n + m = \phi(a^n) + \phi(a^m)$.

CYU 2.24 Let $\phi: G \to G'$ be an isomorphism. For $a', b' \in G'$ let $a, b \in G$ be such that $\phi(a) = a'$ and $\phi(b) = b'$. Then:
$$a'b' = \phi(a)\phi(b) = \phi(ab) = \phi(ba) = \phi(b)\phi(a) = b'a'$$

2.5 Symmetric Groups

CYU 2.25 (a) For $\sigma = \begin{pmatrix} 1 & 2 & 3 & 4 & 5 & 6 & 7 & 8 & 9 & 10 \\ 3 & 9 & 4 & 1 & 5 & 6 & 2 & 7 & 8 & 10 \end{pmatrix}$ we have:

$$\sigma(1) = 3, \sigma^2(1) = \sigma(3) = 4, \sigma^3(1) = \sigma(4) = 1 \Rightarrow (1, 3, 4) \text{ is a cycle.}$$

Picking the first element not moved by the above cycle; namely 2, we have:

$$\sigma(2) = 9, \sigma^2(2) = \sigma(9) = 8, \sigma^3(2) = \sigma(8) = 7, \sigma^4(2) = \sigma(7) = 2 \Rightarrow (2, 9, 8, 7)$$
a cycle

Since the elements not contained in either of the above two cycles are stationary under σ:

$$\sigma = (1, 3, 4)(2, 9, 8, 7)$$

(b) One possible answer: $(1, 2)(3, 4)(5, 6, 7, 8, 9)$.

CYU 2.26 For $\sigma = \begin{pmatrix} 1 & 2 & 3 & 4 & 5 & 6 & 7 & 8 \\ 3 & 8 & 6 & 7 & 4 & 1 & 5 & 2 \end{pmatrix}$ we have:

$$\sigma^2 = \begin{pmatrix} 1 & 2 & 3 & 4 & 5 & 6 & 7 & 8 \\ 6 & 2 & 1 & 5 & 7 & 3 & 4 & 8 \end{pmatrix}, \sigma^3 = \begin{pmatrix} 1 & 2 & 3 & 4 & 5 & 6 & 7 & 8 \\ 1 & 8 & 3 & 4 & 5 & 6 & 7 & 2 \end{pmatrix}, \sigma^4 = \begin{pmatrix} 1 & 2 & 3 & 4 & 5 & 6 & 7 & 8 \\ 3 & 2 & 6 & 7 & 4 & 1 & 5 & 8 \end{pmatrix}$$

$$\sigma^5 = \begin{pmatrix} 1 & 2 & 3 & 4 & 5 & 6 & 7 & 8 \\ 6 & 8 & 1 & 5 & 7 & 3 & 4 & 2 \end{pmatrix}, \sigma^6 = \begin{pmatrix} 1 & 2 & 3 & 4 & 5 & 6 & 7 & 8 \\ 1 & 2 & 3 & 4 & 5 & 6 & 7 & 8 \end{pmatrix}$$ Since 6 is the smallest n for which $\sigma^n = e$, $o(\sigma) = 6$

CYU 2.27 (a) Following the construction above Theorem 2.32 we have:

$$(3, 2, 5, 1) = (3, 1)(3, 5)(3, 2)$$

(b) $\begin{pmatrix} 1 & 2 & 3 & 4 & 5 & 6 & 7 & 8 & 9 & 10 \\ 2 & 4 & 3 & 1 & 10 & 5 & 9 & 6 & 7 & 8 \end{pmatrix} = (1, 2, 4)(5, 10, 8, 6)(7, 9)$

$$= (1, 4)(1, 2)(5, 6)(5, 8)(5, 10)(7, 9)$$

(c) Since $(i, j)(i, j) = (i, j)$, $(i, j)^{-1} = (i, j)$

CYU 2.28 (a) Since $e(i) = i$ for every $1 \leq i \leq n$, e can be expressed as a product of 0 transpositions, and 0 is certainly an even number.

(In the event that $n > 1$, you also have: $e = (1, 2)(1, 2)$)

(b) Since $\begin{pmatrix} 1 & 2 & 3 & 4 & 5 & 6 & 7 & 8 & 9 & 10 \\ 2 & 4 & 3 & 1 & 10 & 5 & 9 & 6 & 7 & 8 \end{pmatrix} = (1, 4)(1, 2)(5, 6)(5, 8)(5, 10)(7, 9)$, the transposition is even.

CYU 2.29 As you can easily check, e, α_1, α_2 are even and the rest are odd. Consequently:

$$A_3 = \langle \{e, \alpha_1, \alpha_2\}, \circ \rangle$$

2.6 Normal Subgroups and Factor Groups

CYU 2.30 We show that the function $f: H \to aH$ given by $f(h) = ah$ is a bijection:

One-to-one: $f(h_1) = f(h_2) \Rightarrow ah_1 = ah_2 \underset{\underset{\text{Theorem 2.10, page 56}}{\uparrow}}{\Rightarrow} h_1 = h_2$.

Onto: For $ah \in aH$, $f(h) = ah$

CYU 2.31 Consider the homomorphism $\phi: \langle Z_2, +_2 \rangle \to S_3$ given by:

$$\phi(0) = \begin{pmatrix} 1 & 2 & 3 \\ 1 & 2 & 3 \end{pmatrix} \text{ and } \phi(1) = \begin{pmatrix} 1 & 2 & 3 \\ 1 & 3 & 2 \end{pmatrix}$$

While Z_2 is normal in Z_2, $\phi(Z_2) = \left\{ \begin{pmatrix} 1 & 2 & 3 \\ 1 & 2 & 3 \end{pmatrix}, \begin{pmatrix} 1 & 2 & 3 \\ 1 & 3 & 2 \end{pmatrix} \right\}$ fails to be a normal subgroup of S_3 (see Example 2.10).

CYU 2.32 (a) Follows from CYU 2.31.

(b) Let $G = \langle a \rangle$ and let $N \triangleleft G$ (actually every subgroup of G is normal). We show that $G/N = \langle aN \rangle$:

Let $gN \in G/N$. Since $g \in \langle a \rangle$, there exists m such that $g = a^m$, We then have:

$$gN = a^m N = \underbrace{aNaN \cdots aN}_{m \text{ times}} = (aN)^m \Rightarrow gN \in \langle aN \rangle$$

The above argument shows that $G/N \subseteq \langle aN \rangle$. Clearly: $\langle aN \rangle \subseteq G/N$.

CYU 2.33 Since, for any $a \in G$, $ag = ga$ for every $g \in G$:

$$Z(G) = \{a \in G | ag = ga \; \forall g \in G\} = G$$

Since for any $a, b \in G$, $aba^{-1}b^{-1} = aa^{-1}bb^{-1} = e$:

$$C(G) = \langle aba^{-1}b^{-1} | a, b \in G \rangle = \langle e \rangle = \{e\}$$

CYU 2.34 We first show that the function $\phi: S_n \to \{-1, 1\}$ given by:

$$\phi(\sigma) = \begin{cases} 1 & \text{if } \sigma \text{ is an even permutation} \\ -1 & \text{if } \sigma \text{ is an odd permutation} \end{cases}$$ is a homomorphism by considering four cases:

If σ and τ are even, then so is $\sigma\tau$, and we have: $\phi(\sigma\tau) = 1 = 1 \cdot 1 = \phi(\sigma)\phi(\tau)$

If σ and τ are odd, then $\sigma\tau$ is even, and we have: $\phi(\sigma\tau) = 1 = (-1)(-1) = \phi(\sigma)\phi(\tau)$

If σ is even and τ is odd, then $\sigma\tau$ is odd, and we have: $\phi(\sigma\tau) = -1 = (1)(-1) = \phi(\sigma)\phi(\tau)$

If σ is odd and τ is even, then $\sigma\tau$ is odd, and we have: $\phi(\sigma\tau) = -1 = (-1)(1) = \phi(\sigma)\phi(\tau)$

Since 1 is the identity in the group $\{-1, 1\}$, $\text{Ker}(\phi) = A_n$. Invoking the First Isomorphism Theorem, we have: $G \cong S_n / A_n$.

CYU 2.35 Employing Theorem 2.42 to the homomorphism $\phi: G \to G'$ we have: $G' \cong G/K$. Restricting ϕ to the group N we arrive at a homomorphism $\phi_N: N \to N'$. In this setting, Theorem 2.42 tels us that $N' \cong N/K$. Consequently: $G/N \cong (G/K)/(N/K)$

2.7 Direct Products

CYU 2.36 (a) **Associativity:**

For $(a_1, a_2, ..., a_n), (b_1, b_2, ..., b_n), (c_1, c_2, ..., c_n) \in G_1 \times G_2 \times \cdots \times G_n$:

$$[(a_1, a_2, ..., a_n)(b_1, b_2, ..., b_n)](c_1, c_2, ..., c_n) = [(a_1 b_1)c_1, (a_2 b_2)c_2, ..., (a_n b_n)c_n]$$
$$= [a_1(b_1 c_1), a_2(b_2 c_2), ..., a_n(b_n c_n)]$$
$$= (a_1, a_2, ..., a_n)[(b_1, b_2, ..., b_n)(c_1, c_2, ..., c_n)]$$

Identity: Letting e_i denote the identity in G_i we have:

$$(a_1, a_2, ..., a_n)(e_1, e_2, ..., e_n) = (a_1 e_1, a_2 e_2, ..., a_n e_n)(a_1, a_2, ..., a_n)$$

Inverses: $(a_1, a_2, ..., a_n)(a_1^{-1}, a_2^{-1}, ..., a_n^{-1}) = (e_1, e_2, ..., e_n)$

(b) If each G_i is abelian, then:

$$(a_1, a_2, ..., a_n)(b_1, b_2, ..., b_n) = (a_1 b_1, a_2 b_2, ..., a_n b_n)$$
$$= (b_1 a_1, b_2 a_2, ..., b_n a_n) = (b_1, b_2, ..., b_n)(a_1, a_2, ..., a_n)$$

Conversely, assume that not all of the G_i are abelian. For definiteness, assume that G_1 is not abelian, with $ab \neq ba$. We then have:

$$(a, e_2, ..., e_n)(b, e_2, ..., e_n) = (ab, e_2, ..., e_n)$$
$$\neq (ba, e_2, ..., e_n) = (b, e_2, ..., e_n)(a, e_2, ..., e_n)$$

CYU 2.37 Noting that 3 has order 2 in Z_6 and is of order 4 in Z_4, and that 4 has order 4 in 16, we conclude that $(3, 3, 4)$ has order $\text{lcm}(2, 4, 4) = 4$ in Z_{30}.

CYU 2.38 Using Induction on s we show that if $n_1, n_2, ..., n_s$ are relatively prime, then the group $Z_{n_1} \times Z_{n_2} \times \cdots \times Z_{n_s}$ is cyclic and isomorphic to $Z_{n_1 n_2 \cdots n_s}$:

I. True if $s = 2$, by Theorem 2.44.

II. Assume True for $s = k$; i.e: $Z_{n_1} \times Z_{n_2} \times \cdots \times Z_{n_k}$ is cyclic and isomorphic to $Z_{n_1 n_2 \cdots n_k}$.

III. We establish validity for $s = k+1$; i.e, that

$Z_{n_1} \times Z_{n_2} \times \cdots \times Z_{n_k} \times Z_{n_{k+1}}$ is cyclic and isomorphic to $Z_{n_1 n_2 \cdots n_k n_{k+1}}$:

$$Z_{n_1} \times Z_{n_2} \times \cdots \times Z_{n_k} \times Z_{n_{k+1}} \cong (Z_{n_1} \times Z_{n_2} \times \cdots \times Z_{n_k}) \times Z_{n_{k+1}}$$

by II: $\cong Z_{n_1 n_2 \cdots n_k} \times Z_{n_{k+1}}$ (with $Z_{n_1 n_2 \cdots n_k}$ cyclic)

by I: $\cong Z_{(n_1 n_2 \cdots n_k) n_{k+1}} \cong Z_{n_1 n_2 \cdots n_k n_{k+1}}$

note that the two number $(n_1 n_2 \cdots n_k)$ and n_{k+1} are relatively prime

In the event that $n_1, n_2, ..., n_s$ are not relatively prime, $Z_{n_1} \times Z_{n_2} \times \cdots \times Z_{n_s}$ is not isomorphic to $Z_{n_1 n_2 \cdots n_s}$ since no element in $Z_{n_1} \times Z_{n_2} \times \cdots \times Z_{n_s}$ has order $n_1 n_2 \cdots n_s$.

CYU 2.39 We are given that $G = HK$ with every $a \in G$ having a unique representation of the form hk. Suppose that $a \in H \cap K$, with $a = hk$. But $a = ae$ is also a representation of a, where $a \in H$ and $e \in K$. It follows, from the unique representation condition, that $k = e$. Similarly, since $a = ea$: $h = e$. Consequently: $a = e$.

CYU 2.40 G_3 has an element of order 8 while G_6 does not.

PART 3
From Rings To Fields

3.1 Definitions and Examples

CYU 3.1 (a) Since $S_3 = \langle S_3, \circ \rangle$ is not an abelian group, it cannot be turned into a ring by imposing any additional operator "$*$".

(b) Let $\langle G, + \rangle$ be an abelian group. By defining $a*b = 0$ for every $a, b \in G$, we arrive at a ring $\langle G, +, * \rangle$.

CYU 3.2 We already know that $\langle nZ, + \rangle$ is an abelian group (Example 2.4, page 62). In addition, nZ is closed under multiplication: $(na)(nb) = n(anb)$. Moreover, since the associative and distributive properties hold for all in integers, they will surely hold for the integers in nZ.

CYU 3.3 Using induction we first show that $n(ab) = (na)b = a(nb)$ for $n \geq 0$:

I. $n(ab) = (na)b = a(nb)$ for $n = 0$.

II. Assume $k(ab) = (ka)b = a(kb)$ for given $k > 0$

III. We show $(k+1)(ab) = [a(k+1)b]$ (A similar argument can be used to show that
$$[(k+1)a]b = [a(k+1)b]):$$
$$(k+1)(ab) = k(ab) + ab \underset{\text{by II}}{=} a(kb) + ab = a(kb+b) = a(k+1)b$$

In the event that $n < 0$ we have:
$$n(ab) = a(nb) \Leftrightarrow -[n(ab)] = -[a(nb)] \underset{\text{by Theorem 3.1(b)}}{\Leftrightarrow} [-n(ab)] \underset{-n > 0}{=} a(-nb)$$

CYU 3.4 (a) We first verify that the nonempty set $M_{2\times 2} = \langle M_{2\times 2}, + \rangle$ satisfies the three properties of Definition 2.1, page 41:

1. $\begin{bmatrix} a_1 & b_1 \\ c_1 & d_1 \end{bmatrix} + \left(\begin{bmatrix} a_2 & b_2 \\ c_2 & d_2 \end{bmatrix} + \begin{bmatrix} a_3 & b_3 \\ c_3 & d_3 \end{bmatrix} \right) = \begin{bmatrix} a_1 & b_1 \\ c_1 & d_1 \end{bmatrix} + \begin{bmatrix} a_2+a_3 & b_2+b_3 \\ c_2+c_3 & d_2+d_3 \end{bmatrix} = \begin{bmatrix} a_1+(a_2+a_3) & b_1+(b_2+b_3) \\ c_1+(c_2+c_3) & d_1+(d_2+d_3) \end{bmatrix}$

$= \begin{bmatrix} (a_1+a_2)+a_3 & (b_1+b_2)+b_3 \\ (c_1+c_2)+c_3 & (d_1+d_2)+d_3 \end{bmatrix}$

$= \left(\begin{bmatrix} a_1 & b_1 \\ c_1 & d_1 \end{bmatrix} + \begin{bmatrix} a_2 & b_2 \\ c_2 & d_2 \end{bmatrix} \right) + \begin{bmatrix} a_3 & b_3 \\ c_3 & d_3 \end{bmatrix}$

2. For every $\begin{bmatrix} a & b \\ c & d \end{bmatrix} \in M_{2\times 2}$: $\begin{bmatrix} a & b \\ c & d \end{bmatrix} + \begin{bmatrix} 0 & 0 \\ 0 & 0 \end{bmatrix} = \begin{bmatrix} a+0 & b+0 \\ c+0 & d+0 \end{bmatrix} = \begin{bmatrix} a & b \\ c & d \end{bmatrix}$

3. For given $\begin{bmatrix} a & b \\ c & d \end{bmatrix} \in M_{2\times 2}$: $\begin{bmatrix} a & b \\ c & d \end{bmatrix} + \begin{bmatrix} -a & -b \\ -c & -d \end{bmatrix} = \begin{bmatrix} a-a & b-b \\ c-c & d-d \end{bmatrix} = \begin{bmatrix} 0 & 0 \\ 0 & 0 \end{bmatrix}$

Moreover, the group $\langle M_{2\times 2}, + \rangle$ is abelian:

$\begin{bmatrix} a_1 & b_1 \\ c_1 & d_1 \end{bmatrix} + \begin{bmatrix} a_2 & b_2 \\ c_2 & d_2 \end{bmatrix} = \begin{bmatrix} a_1+a_2 & b_1+b_2 \\ c_1+c_2 & d_1+d_2 \end{bmatrix} = \begin{bmatrix} a_2+a_1 & b_2+b_1 \\ c_2+c_1 & d_2+d_1 \end{bmatrix} = \begin{bmatrix} a_2 & b_2 \\ c_2 & d_2 \end{bmatrix} + \begin{bmatrix} a_1 & b_1 \\ c_1 & d_1 \end{bmatrix}$

Properties 2 and 3 of Definition 3.1 are also satisfied:

2. $\begin{bmatrix} a_1 & b_1 \\ c_1 & d_1 \end{bmatrix} \left(\begin{bmatrix} a_2 & b_2 \\ c_2 & d_2 \end{bmatrix} \begin{bmatrix} a_3 & b_3 \\ c_3 & d_3 \end{bmatrix} \right) = \begin{bmatrix} a_1 & b_1 \\ c_1 & d_1 \end{bmatrix} \left(\begin{bmatrix} a_2a_3+b_2c_3 & a_2b_3+b_2d_3 \\ c_2a_3+d_2c_3 & c_2b_3+d_2d_3 \end{bmatrix} \right)$

$= \begin{bmatrix} a_1(a_2a_3+b_2c_3)+b_1(c_2a_3+d_2c_3) & a_1(a_2b_3+b_2d_3)+b_1(c_2b_3+d_2d_3) \\ c_1(a_2a_3+b_2c_3)+d_1(c_2a_3+d_2c_3) & c_1(a_2b_3+b_2d_3)+d_1(c_2b_3+d_2d_3) \end{bmatrix}$

$= \begin{bmatrix} (a_1a_2+b_1c_2)a_3+(a_1b_2+b_1d_2)c_3 & (a_1a_2+b_1c_2)b_3+(a_1b_2+b_1d_2)d_3 \\ (c_1a_2+d_1c_2)a_3+(c_1b_2+d_1d_2)c_3 & (c_1a_2+d_1c_2)b_3+(c_1b_2+d_1d_2)d_3 \end{bmatrix}$

$= \left(\begin{bmatrix} a_1 & b_1 \\ c_1 & d_1 \end{bmatrix} \begin{bmatrix} a_2 & b_2 \\ c_2 & d_2 \end{bmatrix} \right) \begin{bmatrix} a_3 & b_3 \\ c_3 & d_3 \end{bmatrix}$

3. $\begin{bmatrix} a_1 & b_1 \\ c_1 & d_1 \end{bmatrix} \left(\begin{bmatrix} a_2 & b_2 \\ c_2 & d_2 \end{bmatrix} + \begin{bmatrix} a_3 & b_3 \\ c_3 & d_3 \end{bmatrix} \right) = \begin{bmatrix} a_1 & b_1 \\ c_1 & d_1 \end{bmatrix} \begin{bmatrix} a_2+a_3 & b_2+b_3 \\ c_2+c_3 & d_2+d_3 \end{bmatrix}$

$= \begin{bmatrix} a_1(a_2+a_3)+b_1(c_2+c_3) & a_1(b_2+b_3)+b_1(d_2+d_3) \\ c_1(a_2+a_3)+d_1(c_2+c_3) & c_1(b_2+b_3)+d_1(d_2+d_3) \end{bmatrix}$

$= \begin{bmatrix} (a_1a_2+b_1c_2)+(a_1a_3+b_1c_3) & (a_1b_2+b_1d_2)+(a_1b_3+b_1d_3) \\ (c_1a_2+d_1c_2)+(c_1a_3+d_1c_3) & (c_1b_2+d_1d_2)+(c_1b_3+d_1d_3) \end{bmatrix}$

$= \begin{bmatrix} a_1 & b_1 \\ c_1 & d_1 \end{bmatrix} \begin{bmatrix} a_2 & b_2 \\ c_2 & d_2 \end{bmatrix} + \begin{bmatrix} a_1 & b_1 \\ c_1 & d_1 \end{bmatrix} \begin{bmatrix} a_3 & b_3 \\ c_3 & d_3 \end{bmatrix}$

In a similar fashion one can show that: $\left(\begin{bmatrix} a_1 & b_1 \\ c_1 & d_1 \end{bmatrix} + \begin{bmatrix} a_2 & b_2 \\ c_2 & d_2 \end{bmatrix} \right) \begin{bmatrix} a_3 & b_3 \\ c_3 & d_3 \end{bmatrix} = \begin{bmatrix} a_1 & b_1 \\ c_1 & d_1 \end{bmatrix} \begin{bmatrix} a_3 & b_3 \\ c_3 & d_3 \end{bmatrix} + \begin{bmatrix} a_2 & b_2 \\ c_2 & d_2 \end{bmatrix} \begin{bmatrix} a_3 & b_3 \\ c_3 & d_3 \end{bmatrix}$

It is easy to show that $\begin{bmatrix} 1 & 0 \\ 0 & 1 \end{bmatrix}$ is the unity in $M_{2 \times 2}$:

$$\begin{bmatrix} 1 & 0 \\ 0 & 1 \end{bmatrix}\begin{bmatrix} a & b \\ c & d \end{bmatrix} = \begin{bmatrix} a & b \\ c & d \end{bmatrix}\begin{bmatrix} 1 & 0 \\ 0 & 1 \end{bmatrix} = \begin{bmatrix} a & b \\ c & d \end{bmatrix}$$

(b) If 1 and $\bar{1}$ are unities in a ring R, then: $1 = (1)(\bar{1}) = \bar{1}$

CYU 3.5 Does there exist $\begin{bmatrix} a & b \\ c & d \end{bmatrix}$ such that $\begin{bmatrix} a & b \\ c & d \end{bmatrix}\begin{bmatrix} 2 & 3 \\ -4 & -6 \end{bmatrix} = \begin{bmatrix} 2a-4b & 3a-6b \\ 2c-4d & 3c-6d \end{bmatrix} = \begin{bmatrix} 1 & 0 \\ 0 & 1 \end{bmatrix}$?

If so, then: $2a - 4b = 1$ and $3a - 6b = 0$, or: $a = \dfrac{1+4b}{2}$ and $a = 2b$, or:

$$\dfrac{1+4b}{2} = 2b \Rightarrow 1 + 4b = 4b \Rightarrow 1 = 0 \,!$$

CYU 3.6 (a) Challenging each element in $Z_6 = \{0, 1, 2, 3, 4, 5\}$ we find that, apart from 1, only 5 has a multiplicative inverses:

multiplying mod 6	2	3	4	5	
2	4	0	2	4	
3	5	3	0	3	
4	2	0	4	2	
5	4	3	2	1	Ah! $5 \cdot 5 = 1$

(b) If m and n are relatively prime then, by Theorem 1.7, page 23: $1 = sm + tn$ for some integers s and t. It follows that $sm = 1 - tn$ which says that sm is congruent to 1 modulo n, and that m is a unit.

If $\gcd(m, n) = d > 1$. Then, by Theorem 1.6, page 22: $d = sm + tn$ for some integers s and t. It follows that $sm = d - tn$ which shows that 1 is not congruent to sm modulo n (it is congruent to d modulo n, with $1 < d \leq n$). It follows that m is not a unit.

CYU 3.7 Expressing Exercise 38 (page 70) in additive form we have:

A (nonempty) subset S of a group G is a subgroup of G if and only if $s, \bar{s} \in S \Rightarrow s - \bar{s} \in S$.

It follows that property (i) of Theorem 3.2: $\langle S, + \rangle$ **is a subgroup of** $\langle R, + \rangle$

can be replace d with: $s, \bar{s} \in S \Rightarrow s\bar{s}^{-1} \in S$.

CYU 3.8 Employing CYU 3.7:

(a) $\begin{bmatrix} 0 & 0 \\ a & b \end{bmatrix} - \begin{bmatrix} 0 & 0 \\ c & d \end{bmatrix} = \begin{bmatrix} 0 & 0 \\ a-c & b-d \end{bmatrix} \in H$ and $\begin{bmatrix} 0 & 0 \\ a & b \end{bmatrix}\begin{bmatrix} 0 & 0 \\ c & d \end{bmatrix} = \begin{bmatrix} 0 & 0 \\ bc & bd \end{bmatrix} \in H$.

(b) Let $x, y \in S_a$. Since $a(x-y) = ax - ay = 0 - 0 = 0$: $x - y \in S_a$.

Since $a(xy) = (ax)y = 0y = 0$: $xy \in S_a$

3.2 Homomorphisms, and Quotient Rings

CYU 3.9 (a) Let $a, b \in \phi^{-1}(H')$. To say that $a - b$ and ab are contained in $\phi^{-1}(H')$ is to say that $\phi(a-b)$ and $\phi(ab)$ are contained in $\phi^{-1}(H')$; and they are:

$$\phi(a-b) = \phi(a) - \phi(b) \in H' \text{ and } \phi(ab) = \phi(a)\phi(b) \in H'$$

(since H' is a subring of G')

(b) Assume there exists an isomorphism $\phi: 3Z \to 5Z$. If so, then:

$$\phi(9) = \phi(3 \cdot 3) = \phi(3)\phi(3) \text{ and } \phi(9) = \phi(3+3+3) = 3\phi(3)$$

This implies that $\phi(3)\phi(3) = 3\phi(3)$ or that $[\phi(3) - 3]\phi(3) = 0$. Since $\phi(3)$ can't be zero (if it were, then ϕ would map everything to zero), $\phi(3)$ must equal 3 — a contradiction since $3 \notin 5Z$.

CYU 3.10 (a) We already know that $I = nZ$ is a subring of Z [Example 3.4(a), page 117]. It is an ideal since, for any $m \in Z$ and $ns \in nZ$: $m(ns) = n(ms) \in nZ$.
The converse follows from the fact that all subgroups of Z are of the form nZ (Exercise 35, page 70).

(b) We already know that $\phi(I)$ is a subring of R' [Theorem 3.4(a)]. It is an ideal:
For $x' \in R'$, choose $x \in R$ such that $\phi(x) = x'$. Then, for any $\phi(a) \in \phi(I)$ we have:

$$x'\phi(a) = \phi(x)\phi(a) = \phi(xa) \in \phi(I) \text{ and } \phi(a)x' = \phi(a)\phi(x) = \phi(ax) \in \phi(I)$$

———— since I is an ideal ————

CYU 3.11 Theorem 2.23(d), page 73, assures us that $I = \varphi^{-1}(I')$ is an additive subgroup of R. As for the second part of Definition 3.6:
Let $i \in I$ and $r \in R$. To show that $ri \in I$ we need but verify that $\phi(ri) \in I'$. Easy enough. Since $\phi(i) \in I'$ and since I' is an ideal in R':

$$\phi(ri) = \phi(r)\phi(i) \in I'$$

A similar argument can be used to show that $ir \in I$.
As $0' \in I'$, $K \subseteq I$. Noting that the function $\phi_I: I \to I'$ given by $\phi_I(i) = \phi(i)$ is an onto homomorphism with kernel K, we have: $I/K \cong I'$.

CYU 3.12 For $a, b \in Z$, let:

(1) $a = q_a n + r_a$, or $r_a = a - q_a n$, where $0 \leq r_a < n$. So: $\phi(a) = r_a$.

(2) $b = q_b n + r_b$, or $r_b = b - q_b n$, where $0 \leq r_b < n$. So: $\phi(b) = r_b$.

(3) $ab = qn + r$, or $r = ab - qn$, where $0 \leq r < n$. So $\phi(ab) = r$.

We compete the proof by showing that $r - r_a r_b \equiv 0 \mod n$:

$$r - r_a r_b \stackrel{3}{=} (ab - qn) - r_a r_b \stackrel{1 \text{ and } 2}{=} ab - qn - (a - q_a n)(b - q_b n)$$

$$= ab - qn - (ab - bq_a n - aq_b n + q_a n q_b n)$$

$$= -qn + bq_a n + aq_b n - q_a n q_b n$$

$$= (-q + bq_a + aq_b + q_a n q_b)n$$

3.3 Integral Domains and Fields

CYU 3.13 (a) $Z_5 = \{0, 1, 2, 3, 4\}$ is a commutative with unity 1. It is easy to see that it has no zero divisors and that every nonzero element has a multiplicative inverse:
$$1 \cdot 1 = 1, \quad 2 \cdot 3 = 3 \cdot 2 = 1, \quad 4 \cdot 4 = 1$$
Conclusion: Z_5 is a field.

(b) $Z_{15} = \{0, 1, 2, \ldots, 14\}$ is a commutative ring with unity 1. It fails to be an integral domain as it has zero divisors ($3 \cdot 5 = 0$).

CYU 3.14 (a) Let R be a commutative ring with unity in which the cancellation property holds. We show that R has no zero divisors (i.e, that R is an integral domain):
Let $ab = 0$. Since $a0 = 0$: $ab = a0$. "Canceling the a," we have $b = 0$.
On the other hand, if R is an integral domain then, by its very definition, it has no zero divisors.

(b) $f_a(x) = f_a(y) \Rightarrow ax = ay \Rightarrow x = y$.

CYU 3.15 By CYU 3.6(b), page 116, every nonzero element in $Z_p = \{0, 1, 2, \ldots, p-1\}$ is a unit. It follows that Z_p is an integral domain, and therefore a field (Theorem 3.12).

CYU 3.16 (a) Consider the factored form $x^2 - x - 6 = (x-3)(x+2) = 0$. Clearly $x = 3$ is a solution. Not quite as clear is that 10 is also a solution, as $10 + 2 = 0$. From the discussion preceding this CYU we know that there are five pairs of numbers in $Z_{12} = \{0, 1, 2, 3, 4, 5, 6, 7, 8, 9, 10, 11\}$ (not involving 0) whose product equals 0 (in Z_{12})— specifically: $(2, 6), (3, 4), (8, 3), (9, 4)$ and $(10, 6)$. The plan, now, is to find those elements x in Z_{12} for which the product $(x-3)(x+2)$ turns out to involve any or the above five pairs. A direct calculation shows that 6 and 7 are the only winners:
$$\text{For } x = 6: (x-3)(x+2) = (6-3)(6+2) = 3 \cdot 8$$
$$\text{For } x = 7: (x-3)(x+2) = (7-3)(7+2) = 4 \cdot 9$$
Conclusion: 3, 10, 6, and 7 are the solutions of $x^2 - x - 6 = 0$ in Z_{12}.

(b) In $Z_8 = \{0, 1, 2, 3, 4, 5, 6, 7\}$ the equation $x^2 - x - 6 = (x-3)(x+2) = 0$ is seen to have solutions 3 and 6 (since $6 + 2 = 0$). Here are the only pairs (not involving 0) with product equal to 0 (in Z_8): $(2, 4)$ and $(4, 6)$. A direct calculation shows that for no x in Z_8 does the product $(x-3)(x+2)$ involve either $(2, 4)$ or $(4, 6)$. For example, if $x = 5$, then $(x-3)(x+2)$ involves the pair $(2, 7)$.

Conclusion: 3 and 6 are the solutions of $x^2 - x - 6 = 0$ in Z_8.

CYU 3.17 We first acknowledge the fact that the ring $P_{Z_n}[x]$ is infinite, for it contains the infinite set of polynomials $\{x^n\}_{n=1}^{\infty}$. As the coefficients of any polynomial $p(x)$ in $P_{Z_n}[x]$ are elements in Z_n, $np(x) = 0$ for every $p(x) \in P_{Z_n}[x]$. Moreover, for no $0 < m < n$ is it true that $np(x) = 0$, where $p(x)$ is the constant polynomial 1. It follows that $P_{Z_n}[x]$ has characteristic n.

CYU 3.18 Expanding $(a+b)^3$ we find that $(a+b)^3 = a^3 + 3a^2b + 3ab^2 + b^3$. Since the given domain D had characteristic 3, the terms $3a^2b$ and $3ab^2$ are zero. Consequently: $(a+b)^3 = a^3 + b^3$.

CYU 3.19 (a) If pZ is not a maximal ideal, then $pZ \subset nZ$ for some $n > 1$ [see CYU 3.10(a), page 124]. Consequently $p = nm$ for some $m \in Z$ — contradicting the given condition that p is prime.

(b) Assume that $I = mZ$ is a maximal ideal. If m is not prime, then $m = ab$ with neither a nor b equal to 1. But then:

$$mZ \subseteq aZ \subsetneq Z$$
$$\uparrow \qquad \uparrow$$
$$mk \in aZ \; \forall k \in Z \text{ since } mk = a(bk) \qquad 1 \notin aZ$$

Contradicting the given condition that mZ is a maximal ideal in Z.

So, every maximal ideal in Z is of the form pZ for p prime. As such, it is a prime ideal, for:

$$ab \in pZ \Rightarrow p|ab \underset{\uparrow}{\Rightarrow} p|a \text{ or } b|b \Rightarrow a \in pZ \text{ or } b \in pZ$$
$$\text{Theorem 1.9, page 24}$$

Conversely, assume that $I = mZ$ is a prime ideal in Z. If m is not prime, then $m = ab$ with neither a nor b equal to 1. But then I is not a prime ideal, since $ab \in mZ$ with neither a nor b contained in mZ. So:

$$I = mZ \text{ prime} \Rightarrow m \text{ prime} \underset{\uparrow}{\Rightarrow} I = mZ \text{ maximal}$$
$$\text{(a)}$$

APPENDIX B

We offer Professor Goldberg's proof that the groups Z_4 and K appearing in Figure 2.1, page 43, are the only groups of order 4.

PROPOSITION: Let S be a group of order 4, with identity e. Then, for every $a \in S$, there exists a positive integer d so that:
- $a^d = e$;
- $a^k \neq e$, for any positive integer k smaller than d; and
- $d = 1, 2,$ or 4.

PROOF: Choose an arbitrary element a of S. Consider the following elements of S: e, a, a^2, a^3. Since S has 4 elements, by an elementary application of the pigeonhole principle, either:

case 1. $a^i = e$, $i = 1, 2, 3,$ or 4 and/or

case 2. $a^i = a^j$, some integers i, j, $1 \leq j < i \leq 4$.

In either case (using inverses for case 2), we obtain that $a^k = e$ for some integer $k \in \{1, 2, 3, 4\}$:

for case 1, $k = i$; for case 2 $k = i - j$.

Hence the set $\{k \in Z^+ | a^k = e\}$ is not empty. Let d be the minimum of this set. From the above, $1 \leq d \leq 4$.

To finish the proof of the Proposition, it remains to show that d cannot be 3:

If $d = 3$, the elements e, a, a^2 are distinct. Since S is of order 4, $\exists b \in S$ with $b \notin \{e, a, a^2\}$. So, $\{e, a, a^2, b\} = S$. By closure, $ab \in S$. But note that:

$ab \neq b$, since $ab = b \Rightarrow a = e$, impossible;

$ab \neq a^2$, since $ab = a^2 \Rightarrow b = a$, impossible;

$ab \neq a$, since $ab = a \Rightarrow b = e$, impossible;

$ab \neq e$, since $ab = e \Rightarrow a^2(ab) = a^2$ (using that $a^3 = e$), impossible.

Hence S contains at least 5 distinct elements, contradicting that it has order 4. So d cannot be 3.

Using the above Proposition, it is easy to classify groups of order 4. By the Proposition, there are 2 cases:

case 1. $\exists a \in S$ with $a \neq e$, $a^2 \neq e$, $a^3 \neq e$, and $a^4 \neq e$, or

case 2. $\forall a \in S$, $a^2 = e$.

In Case 1, we have $S = \{e, a, a^2, a^3\}$ and $a^4 = e$ Up to "renaming", S is Z_4. In Case 2, pick an element $a \in S$ with $a \neq e$. Next, pick an element $b \neq e$ with $b \neq a$. Since we are in Case 2, $a^2 = b^2 = e$, and $e = (ab)^2 = abab$. Multiplying on the left by a and on the right by b we obtain $ab = ba$. Hence, $S = \{e, a, b, ab\}$ (it is easy to see that ab does not equal $e, a,$ or b), each element is of order 2, and S is commutative.

DETERMINANTS

We define a function that assigns to each square matrix a (real) number:

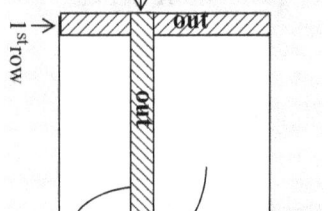

resulting in a square matrix of dimension $n-1$

DETERMINANT

For $A = \begin{bmatrix} a & b \\ c & d \end{bmatrix}$:

$$\det \begin{bmatrix} a & b \\ c & d \end{bmatrix} = ad - bc$$

For $A \in M_{n \times n}$, with $n > 2$, let A_{1j} denote the $(n-1) \times (n-1)$ matrix obtained by deleting the first row and j^{th} column of the matrix A (see margin). Then:

$$\det(A) = \sum_{j=1}^{n} (-1)^{1+j} a_{1j} \det(A_{1j})$$

The above definition defines the determinant of a matrix by an expansion process involving the first row of the given matrix. The following theorem (proof omitted), known as the Laplace Expansion Theorem, enables one to expand along any row or column of the matrix.

Note that the sign of the $(-1)^{i+j}$ has an alternating checkerboard pattern

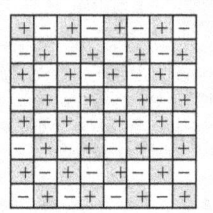

THEOREM 1

For given $A \in M_{n \times n}$, A_{ij} will denote the $(n-1) \times (n-1)$ submatrix of A obtained by deleting the ith row and j^{th} column of A. We then have:

EXPANDING ALONG THE ith ROW

$$\det(A) = \sum_{j=1}^{n} (-1)^{i+j} a_{ij} \det(A_{ij})$$

and:

EXPANDING ALONG THE jth COLUMN

$$\det(A) = \sum_{i=1}^{n} (-1)^{i+j} a_{ij} \det(A_{ij})$$

THEOREM 2

The determinant of the identity matrix $I_n \in M_{n \times n}$ is 1.

PROOF: By induction on the dimension, n, of $M_{n \times n}$.

I. Holds for $n = 2$: $\det \begin{bmatrix} 1 & 0 \\ 0 & 1 \end{bmatrix} = 1$.

II. Assume claim holds for $n = k$: $\det(I_k) = 1$

III. We establish validity at $n = k+1$: $\det(I_{k+1}) = 1$

Expanding across the first row of I_{k+1} (see margin) we have:
$$\det(I_{k+1}) = \det(I_k) = 1$$

THEOREM 3   If two rows of $A \in M_{n \times n}$ are interchanged, then the determinant of the resulting matrix is $-\det(A)$.

PROOF: By induction on the dimension of the matrix A. For $n = 2$:
$$\det\begin{bmatrix} a & b \\ c & d \end{bmatrix} = ad - bc \quad \text{and} \quad \det\begin{bmatrix} c & d \\ a & b \end{bmatrix} = cb - da$$

— negative of each other —

Assume the claim holds for matrices of dimension $k > 2$ (the induction hypothesis).

Let $A = [a_{ij}]$ be a matrix of dimension $k+1$, and let $B = [b_{ij}]$ denote the matrix obtained by interchanging rows p and q of A. Let i be the index of a row **other than** p and q. Expanding about row i we have:

$$\det(A) = \sum_{j=1}^{k+1} (-1)^{i+j} a_{ij} \det(A_{ij}) \quad \text{and} \quad \det B = \sum_{j=1}^{k+1} (-1)^{i+j} b_{ij} \det(B_{ij})$$

Since rows p and q were switched to go from A to B, row i of B still equals that of A, and therefore: $b_{ij} = a_{ij}$. Since B_{ij} is the matrix A_{ij} with two of its rows interchanged, and since those matrices are of dimension k, we have: $\det \mathbf{B}_{ij} = -\det A_{ij}$ (the induction hypothesis).

Consequently:

$$\det(\mathbf{B}) = \sum_{j=1}^{k+1} (-1)^{i+j} b_{ij} \det(\mathbf{B}_{ij}) = \sum_{j=1}^{k+1} (-1)^{i+j} a_{ij} [-\det(A_{ij})]$$

$$= -\sum_{j=1}^{k+1} (-1)^{i+j} a_{ij} \det(A_{ij}) = -\mathbf{det}(\mathbf{A})$$

B-4 Determinants

Appendix C
Answers to Selected Exercises
Part 1
Preliminaries

1.1 Functions.

1. U **3.** $\{15n \mid n \in U\}$ **5.** B **7.** D **9.** U **11.** $\varnothing$ **13.** F **15.** D
17. $\{1, 3, 5, 7, 9, 11, 13, 15\}$ **19.** E **27.** One-to-one and onto
29. Not one-to-one, not onto   **31.** One-to-one not onto   **33.** Not one-to-one, not onto
35. One-to-one and onto   **37.** Not one-to-one, not onto   **39.** Not one-to-one, not onto

41. $f^{-1}(y) = \dfrac{y+2}{3}$ **43.** $f^{-1}(y) = \dfrac{y}{2-y}$ **45.** $f^{-1}(a, b) = \left(\dfrac{a}{5}, b-3\right)$

47. $f^{-1}\begin{bmatrix} a & b \\ c & d \end{bmatrix} = \begin{bmatrix} c & -d \\ a & \dfrac{b}{2} \end{bmatrix}$ **49.** $f^{-1}(a, b, c) = \begin{bmatrix} \dfrac{a}{2} \\ \dfrac{1}{2}(s-2b) \\ \dfrac{1}{2}(-a+2b-2c) \end{bmatrix}$

1.2 Principle of Mathematical Induction.

Each exercise calls for a verification or proof.

1.3 The Division Algorithm and Beyond.

1. $q = r = 0$ **3.** $q = -27, r = 1$ **5.** 5   **7.** 60   **9.** 90

1.4 Equivalence Relations.

23. Yes   **25.** No   **27.** Yes   **29.** No   **31.** Yes   **33.** Yes

43. $[n] = \{n + 5k \mid k \in Z\}$ **45.** $[x] = \{-x, x\}$ **47.** $[(x_0, y_0)] = \{(x_0, y) \mid (y \in \Re)\}$

Part 2
Groups

2.1 Definitions and Examples.

1. A cyclic group with generator 2.   **2.** Not a group. It does not contain an identity.

3. Not a group. It does not contain an identity.   **5.** Not a group. It does not contain an identity.

7. Not a group. 1 is the identity, but 2 has no inverse.   **9.** Abelian group. Not cyclic.

11. Abelian group. Not cyclic.

13. $\alpha_3^2 = e$, $\alpha_3^3 = \alpha_3$ **15.** $\alpha_1^n = \begin{cases} e & \text{if } n \equiv 0 \bmod 3 \\ \alpha_1 & \text{if } n \equiv 1 \bmod 3 \\ \alpha_2 & \text{if } n \equiv 2 \bmod 3 \end{cases}$ **17.** $\alpha_3^{-n} = \begin{cases} e & \text{if } n \text{ is even} \\ \alpha_3 & \text{if } n \text{ is odd} \end{cases}$

19. $\alpha_2^2 = \alpha_1$, $\alpha_2^3 = e$ **21.** $\alpha_2^n = \begin{cases} e & \text{if } n \equiv 0 \bmod 3 \\ \alpha_2 & \text{if } n \equiv 1 \bmod 3 \\ \alpha_1 & \text{if } n \equiv 2 \bmod 3 \end{cases}$ **23.** $\alpha_2^{-n} = \begin{cases} e & \text{if } n \equiv 0 \bmod 3 \\ \alpha_1 & \text{if } n \equiv 1 \bmod 3 \\ \alpha_2 & \text{if } n \equiv 2 \bmod 3 \end{cases}$

25. $\beta\alpha = \begin{pmatrix} 1 & 2 & 3 & 4 & 5 & 6 \\ 1 & 4 & 3 & 6 & 5 & 2 \end{pmatrix}$ **27.** $\gamma\beta = \begin{pmatrix} 1 & 2 & 3 & 4 & 5 & 6 \\ 5 & 6 & 3 & 4 & 1 & 2 \end{pmatrix}$ **29.** $\alpha^5 = \alpha^{-1} = \begin{pmatrix} 1 & 2 & 3 & 4 & 5 & 6 \\ 6 & 1 & 2 & 3 & 4 & 5 \end{pmatrix}$

31. $\alpha^{101} = \alpha^5 = \alpha^{-1} = \begin{pmatrix} 1 & 2 & 3 & 4 & 5 & 6 \\ 6 & 1 & 2 & 3 & 4 & 5 \end{pmatrix}$ **33.** $\beta^{101} = \beta^5 = \beta^{-1} = \begin{pmatrix} 1 & 2 & 3 & 4 & 5 & 6 \\ 2 & 1 & 4 & 3 & 6 & 5 \end{pmatrix}$

35. An abelian non-cyclic group.   **37.** Not a group   **39.** A cyclic abelian group

2.2 Elementary Properties of Groups.

1. (a) e (b) a (c) $a^{-1}cb^{-1}$ (d) aba^{-1}

2.3 Subgroups.

1. Yes   **3.** No   **5.** Yes   **7.** No   **9.** Yes   **11.** Yes   **13.** No   **15.** Yes   **17.** No
19. Yes   **21.** Yes   **23.** Yes   **25.** No   **27.** Yes   **29.** No   **31.** Yes   **33.** Yes

2.4 Homomorphisms and Isomorphisms.

1. Yes   **3.** No   **5.** Yes   **7.** Yes   **9.** Yes   **11.** Yes   **13.** Yes

2.5 Symmetric Groups.

1. $(2,5)(1,3,4) = (2,5)(1,4)(1,3)$ **3.** $(1,5,2) = (1,2)(1,5)$

5. $(2,5,6,4) = (2,4)(2,6)(2,5)$ **7.** $(1,6,3,4)(2,5) = (1,4)(1,3)(1,6)(2,5)$

9. $(1,2)(3,4)$ **11.** 3 **13.** 4 **15.** 7 **17.** $\sigma = (1,2,4,5,3)$ **19.** $\sigma = (1,3,4,2)$

2.6 Normal Subgroups and Factor Groups

1. No **3.** Yes

2.7 Direct Products.

1. 36 **3.** 36 **5.** 36

7. Order 1: $(0,0)$, order 2: $(1,0)$, order 3: $(0,1)$, $(0,2)$, order 4: $(1,1)$, $(1,2)$

9. The element $(0,0,e)$ has order 1. The remaining seven elements have order 2.

11. $\{0,0\}$, $\{(0,0),(0,1),(0,2)\}$

13. Here are the proper subgroups of $Z_2 \times Z_2 \times S_2$, where $S_2 = \{e, \sigma\}$ with $\sigma = \begin{pmatrix} 1 & 2 \\ 2 & 1 \end{pmatrix}$:

$\{(0,0,e)\}$, $\{(0,0,e),(1,0,e)\}$, $\{(0,0,e),(0,1,e)\}$, $\{(0,0,e),(0,0,\sigma)\}$

$\{(0,0,e),(1,1,e)\}$, $\{(0,0,e),(0,1,\sigma)\}$, $\{(0,0,e),(1,0,\sigma)\}$, $\{(0,0,e),(1,1,\sigma)\}$

$\{(0,0,e),(0,1,e),(1,0,\sigma),(1,1,\sigma)\}$, $\{(0,0,e),(1,0,e),(0,1,\sigma),(1,1,\sigma)\}$

$\{(0,0,e),(0,0,\sigma),(1,1,e),(1,1,\sigma)\}$

15. $Z_2 \times Z_2 \times Z_3 \times Z_3$ **17.** $Z_2 \times Z_2 \times Z_5 \times Z_3 \times Z_3$ **19.** 10

$Z_4 \times Z_2 \times Z_3$ $Z_4 \times Z_5 \times Z_3 \times Z_3$

$Z_2 \times Z_2 \times Z_9$ $Z_2 \times Z_2 \times Z_5 \times Z_9$

$Z_4 \times Z_9$ $Z_4 \times Z_5 \times Z_9$

Part 3
From Rings to Fields

3.1 Definitions and Examples.

1. Yes **3.** No **5.** Yes **7.** Yes **9.** Yes **11.** No **13.** No **15.** Yes

17. No **19.** Yes **21.** –1, 1 **23.** 1, 2, 3, 4 **25.** (1, 1), (–1, 1), (1, –1), (–1, –1)

27. (1, 1), (1, 2), (1, 4), (1, 5)(1, 7), (1, 8), (5, 1), (5, 2), (5, 4), (5, 5)(5, 7), (5, 8)

3.2 Homomorphisms and Quotient Rings.

1. Yes **3.** No **5.** No **7.** No **9.** Yes

19. $\phi(n) = (n, n)$ is the only homomorphism from Z to $Z \times Z$.

3.3 Integral Domains and Fields.

1. None **3.** (0, 2), (0, 3) **5.** None **7.** 0 **9.** 6 **11.** 10

A

Abelian Group, 43
Addition Modulo n, 42
Alternating Group, 88
Alternate Principle of Induction, 17
Automorphism, 76

B

Bijection, 4

C

Cardinality, 31
Cancellation Law 55
Cartesian Product, 2, 103
Cayley's Theorem, 78
Center Subgroup, 96
Characteristic, 131
Commutative Ring, 111
Commutator Subgroup, 96
Composition, 3
Congruence Modulo n, 33
Cycle, 84
    Decomposition, 84
Cyclic Group, 48
    Generator, 48

D

Direct Product, 103
    External, 103
    Internal, 105
Division Algorithm, 21
Divisibility, 15
Domain, 2

E

Equivalence Class, 31
Equivalence Relation, 29
Even Integer, 15
Even Permutation, 88

F

Factor Group, 95
Field, 128
    of Quotients, 133

Function, 2
    Bijection, 4
    Domain, 2
    Composition, 3
    Inverse, 5
    One-to-One, 4
    Onto, 4
    Range, 2
Fundamental Theorem of
    Finitely Generated Abelian Groups, 107
Fundamental Counting Principle, 107

G

Generator, 48
Generated Subgroup, 65
Greatest Common Divisor, 22
Group, 41
    Abelian, 43
    Alternating, 88
    Cyclic, 48
        Generator, 48
    Factor, 95
    Klein, 43
    Order, 43
    Symmetric, 84
    Table, 43

H

Homomorphism (Group), 72
    Image, 75
    Kernel, 75
Homomorphism (Ring), 121

I

Ideal, 123
    Maximal, 132
    Prime, 132
Image, 75
Index of a subgroup, 93
Induction, 13
Integral Domain, 128
Inverse Function, 5
Isomorphism (Group), 75
    Theorems, 98
        First, 98
        Second, 99
        Third, 99
Isomorphism (Ring), 116
    Theorems, 124
        First, 124

Second, 125
Third, 125
Inverse Function, 5

K

Kernel, 74
Klein Group, 43

L

Lagrange's Theorem, 65
Least Common Multiple, 104

M

Mathematical Induction 13, 16
Maximal Ideal, 132

N

Normal Subgroup, 93

O

Odd Integer, 15
Odd permutation, 88
One-to-One, 4
Onto, 4
Order of an Element, 58
Order of a Group, 43
Order of a Permutation, 86

P

Partition, 32
Permutation, 46
    Even, 88
    Odd, 88
Prime number, 24
Prime Ideal, 132
Principle of Mathematical Induction, 13,16

Q

Quotient Group, 95
Quotient Field, 133

Quotient Ring, 123

R

Range, 2
Relation, 29
    Equivalence, 29
    Reflexive, 29
    Symmetric, 29
Relatively Prime, 23
Ring, 111
    characteristic, 131
    Commutative, 113
    With Unity, 113

S

Set, 1
    Complement, 1
    Disjoint, 1
    Equality, 1
    Intersection, 1
    Subset, 1
        Proper, 1
    Union, 1
Subgroup, 62
    Center, 96
    Commutator, 96
    Generated by a, 63
    Generated by subsets, 65
    Normal, 93
Subring, 116
Symmetric Group, 46, 84

T

Transposition, 85

U

Unit, 114
Unity, 113

W

Well Ordering Principle, 18

Z

Zero Divisor, 128

www.ingramcontent.com/pod-product-compliance
Lightning Source LLC
Chambersburg PA
CBHW082329220526

45470CB00008B/2448